Hi-Tech Horticulture

Volume 2: Organic Farming and Sustainable Development Techniques

OTHER VOLUMES OF SERIES

ISBN 9789387973404
INR 2700.00

ISBN 9789387973428
INR 2700.00

ISBN 9789387973435
INR 2700.00

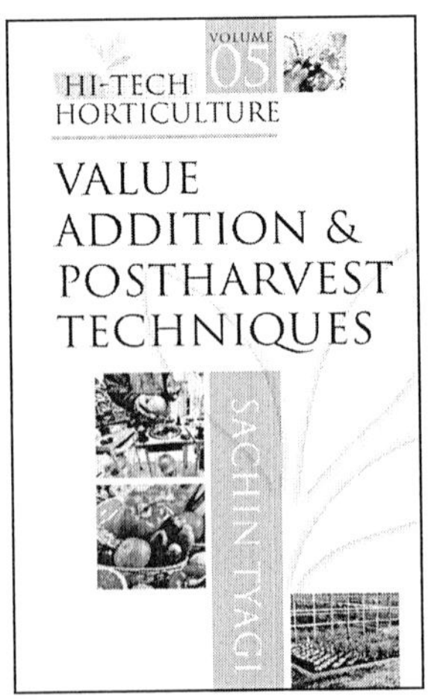

ISBN 9789387973459
INR 2700.00

ISBN 9789387973466
INR 2700.00

Hi-Tech Horticulture

Volume 2: Organic Farming and Sustainable Development Techniques

Sachin Tyagi
Department of Horticulture (F & FT)
Bihar Agriculture University
Sabour, Bhagalpur-813210 (Bihar) India

NEW INDIA PUBLISHING AGENCY
New Delhi – 110 034

NEW INDIA PUBLISHING AGENCY
101, Vikas Surya Plaza, CU Block, LSC Market
Pitam Pura, New Delhi 110 034, India
Phone: + 91 (11) 27 34 17 17 Fax: + 91 (11) 27 34 16 16
Email: info@nipabooks.com
Web: www.nipabooks.com

Feedback at feedbacks@nipabooks.com

ISBN: 978-93-87973-41-1

Composed, Designed and Printed in India

Preface to the Series

Horticulture plays an important part in today's agriculture and there are new avenues that are being achieved by horticulture. The subject has transformed from only being about vegetables, fruits, flowers and postharvest technology and has moved towards disease, breeding, pathology, physiology, greenhouse technologies and other areas which were never heard for.

The book series: Hi-Tech Horticulture has been worked out keeping the above mentioned issues in mind with contribution by eminent professors and scientists. The volumes are on:

1. Crop Improvement Nursery and Rootstock Management
2. Organic Farming and Sustainable Development Techniques
3. Improved Production Techniques
4. Plant Protection and Stress Management
5. Value Addition and Post-Harvest Management
6. Advance Techniques

Series Editor

Preface to the Volume

The present volume has been contributed by eminent professors and scientist in their field and updated and most recent developments have been covered with.

I acknowledge the support from my university colleagues and contributors.

Encouragements by the present Vice Chancellor Dr Ajoy Kumar Singh is duly acknowledged.

Sachin Tyagi

Contents

Section 01
Organic Farming

1

Recent Techniques of Organic Farming or Eco-Friendly Horticulture

Rahul Kumar and Mamta Singh

Crop Improvement Division, ICAR-Indian Institute of Vegetable Research Varanasi- 221 305 (Uttar Pradesh) India

Introduction : Need to Change

During the era of Green Revolution, India remarkably achieved self-sufficiency in food production with the introduction of high-yielding varieties, extension in the irrigated cultivable land, high use of chemical fertilizer such as Urea and other sources of NPK fertilizers and increase in cropping intensity.During these processes, relative contribution of organic manures as a source of plant nutrients declined substantially. Due to heavy application of chemical pesticides an increase in resistance of insect pests has also been noticed for most of the regular used plant protection chemicals. Health hazards associated with intensive modern agriculture, such as pesticides residues in food products and groundwater contamination are matter of concern. In the progress of green revolution technologies and chemical dependent nutrient management has declined the overall productive capacity of the soil due to no judicious fertilizer use. Such concerns and problems faced by modern-day agriculture gave birth to new concepts in farming, such as organic farming, natural farming, biodynamic agriculture, donothing agriculture, eco-farming, etc. The essential feature of such farming practices imply, i.e., back to nature.

Adverse impact of Green Revolution (1967-1977) at a glance

1. Soil fatigue due to intensive cultivation.
2. Stagnation of HYV yield.
3. Decrease in input use efficiency.
4. Declining water table.

5. Increase susceptibility to pest & diseases.
6. Pest becoming tolerant to pesticides.
7. Increased soil salinity.
8. Serious imbalance in nutrient status.
9. Deficiency in secondary & micronutrients i.e. Sulphur, Zinc, Boron, Iron, Manganese, Molybdenum including universal deficiency of NPK.
10. Nitrate contamination in ground water.
11. Accumulation of heavy metals like Arsenic, Lead & Cadmium.
12. Presence of pesticide residue in food material and milk sample.

Organic farming has the twin objective of the system, sustainable and environmentally sensitive. In order to achieve these two goals, it has developed some rules and standards which must be strictly adhere to. There is very little scope for change and flexibility. Thus, the Organic Farming does not require best use of options available rather the best use of options that have been approved. These options are usually more complex and less effective than the conventional system. With ever increasing population having huge requirements of vegetables and meagre availability of organic resources, pure organic farming is not possible in India; rather some specific area can be diverted to organic farming for export of high quality vegetable crops. Thus, as a whole under Indian condition, only partial switching to Organic Farming of export oriented vegetable crops can be possible in recent times. In this context, it is relevant to quote Nobel Laureate Dr. Norman Borlaug (2002) who said that," Switching on food production to organic would lower crop yields. We can use all the organic that are available but we are not going to feed six billion people with organic fertilizers."

The use of word "organic" has been popularized by Howard and Rodale and the term 'organic' was first used in relation to farming by Northbourne (1940) in his book Look to the Lad.Biodynamic agriculture was the first modern system of agriculture to focus exclusively on organic methods (Paull and John, 2011). Its development began in 1924 onwards and in support of Biodynamic agriculture Rudolf Steiner entertained a series of eight lectures on agriculture (Paull and John, 2013). These lectures, the first known presentation of what later came to be known as organic agriculture, were held in response to a request by farmers who noticed degraded soil conditions and a deterioration in the health and quality of crops and livestock resulting from the use of chemical fertilizers. The one hundred eleven attendees, less than half of whom were farmers, came from six countries, primarily Germany and Poland. The lectures were published in November 1924; the first English translation appeared in 1928 as The Agriculture Course (Paull and John, 2011).

Definitions of Organic Farming

Many scientists at different levels have elaborated the concept of Organic Farming; the important descriptions are as follows :

Organic agricultural methods are internationally regulated and legally enforced by many nations, based in large part on the standards set by the International Federation of Organic Agriculture Movements (IFOAM), an international umbrella organization for organic farming organizations established in 1972 (Paull and John, 2010). Organic agriculture can be defined as: An integrated farming system that strives for sustainability, the enhancement of soil fertility and biological diversity whilst, with rare exceptions, prohibiting synthetic pesticides, antibiotics, synthetic fertilizers, genetically modified organisms, and growth hormones (D. Treadwell *et al.*, H. Martin and D. Rhoads).

Organic agriculture is a holistic production management system which promotes and enhances agro-ecosystem health, including biodiversity, biological cycles, and soil biological activity. It emphasizes, the use of management practices in preference to the use of off –farm inputs, taking into account that regional conditions require locally adapted systems. This is accomplished by using, where possible, agronomic, biological, and mechanical methods, as opposed to using synthetic materials, to fulfill any specific function within the system (FAO, 1999).

Description offered by Lampkin (1990) has been found to be the most comprehensive one covering all essential features of Organic Farming. According to Lampkin 'Organic Farming is a production system which avoids or largely excludes the use of synthetically compounded fertilizers, growth regulators and live stock feed additives. To the maximum extent, feasible Organic Farming systems rely on crop rotations, crop residues, animal manures, legume. Green manures, off farm organic wastes and aspects of biological pest control to maintain soil productivity and tilt to supply plant nutrients and to control to maintain soil productivity and tilth to supply plant nutrients and to control insect pests, diseases and weeds. Thus, the Organic farming implies recycling of waste and residue to the native soil itself, replenishing the nutrients depleted from the soil during the crop growth, encouraging the growth of microorganisms which could regulate phased release of soil moisture and soil aeration and ensuring soil fertility by firmly binding the nutrients element in the complex organic molecules.

Among the most stringent definitions is that of US Department of. Agriculture, which Organic Farming as, 'A system that is designed and mailed to produce to agricultural products by the use of methods, and substances that maintain the integrity of organic agricultural products until consumer". This is accomplished by using, where possible, cultural, biological and mechanical methods, as oppose to use substances to fulfill any specific fluctuations within the system so as to maintain long activity, ensure effective management, recycle waste to return nutrients to the land, provides attentive cares for farm animals and handle the agricultural products without the use of extraneous synthetic additives or processing in accordance with the act and regulations in this part (USDA, 1980)

According to Funtilana (1990), "Organic Farming is giving back to the nature what is taken from it". It is not mere non-chemicalism in agriculture; it is a system of farming based on integral relationship. Therefore, one should know the relationship among soil, water, plant and micro flora and overall relationship between plants animal kingdom. It is the totality of these relationships, which is the backbone of the Organic Farming. Above all, the success of organic farming depends to a great extent on the efficiency of agronomic management adopted to stimulate and augment the underlying productivity of the soil resources. All the managemental practices followed in organic farming are governed by the principles of ecology and are within the ecological means.

Indian Initiatives

Since when agriculture came to existence in India, farming relied on organic-like practices. Even presently, much of the forest produce of economic importance, such as herbs, medicinal plants, etc., is in this category. As per the available statistics, India's rank in terms of World's Organic Agricultural land was 15, up to the year 2013 (Source FIBL & IFOAM Year Book 2015). The total area under organic certification is 5.71million Hectare (2015-16). This includes 26% cultivable area with 1.49 million Hectare and rest 74% (4.22 million Hectare) forest and wild area for collection of minor forest produces. India produced around 1.35 million MT (2015-16) of certified organic products which includes all varieties of food products namely Sugarcane, Oil Seeds, Cereals & Millets, Cotton, Pulses, Medicinal Plants, Tea, Fruits, Spices, Dry Fruits, Vegetables, Coffee etc. . The production is not limited to the edible sector but also produces organic cotton fiber, functional food products etc. The total volume of export during 2015-16 was 263687 MT. The organic food export realization was around 298 million USD.

India has brought 4.72 million ha under organic certification processes, including 0.6 million ha of cultivated agricultural land and 4.12 million ha for wild harvest collection in forests, as of March 2014 (A.K.Yadav, 2015). Among all the states, Madhya Pradesh has covered largest area under organic certification followed by Himachal Pradesh and Rajasthan. Sikkim has become India's first fully organic state by implementing organic practices on approximately 75,000 ha of agricultural land (Anno. 2016). There has been a consistent increase in the number of certified farmers in the country every year. Organic farmers' associations have taken the lead in adopting and spreading technology. Institutional support for organic production was created by the launch of the National Programme for Organic Production (NPOP) by the Agriculture and Processed Food Export Development Authority (APEDA), Ministry of Commerce. The NPOP supports promotional initiatives, accreditation of inspection and certification agencies, and offers support to agri-business enterprises to facilitate export. APEDA has been interacting with the European Union (EU), the United States Department of Agriculture (USDA), Japan, and IFOAM for recognition of equivalence of the Indian quality assurance system. India was recognized for the purpose of equivalence since June 2006 for an unspecified duration, as reported in the Annex III of Regulation (EC) 1235/2008. However, the European Union's organic production rules are not formally recognized as equivalent by India.

In 2004, India's Ministry of Agriculture launched a nationwide initiative—the National Project on Organic Farming (NPOF) with the objective of promoting organic farming in India by facilitating access to organic inputs, streamlining production, facilitating certiûcation, and developing domestic markets for organic commodities. In 2004–2005,a national Centre for Organic Farming was established under the Ministry of Agriculture at Ghaziabad to provide institutional support and to facilitate moving farmers in to organic crop production by providing suitable logistics. The National Horticulture Mission launched by India's Department of Agriculture and Cooperation in 2005 offers assistance for transitioning to organic farming of horticultural crops. As a result of these interventions, organic agriculture has seen unexpectedly high growth. By the combined effort of farmers, the government, NGOs, and market forces, the Indian organic movement has reached a stage where it can swiftly move to occupy a desired space in Indian agriculture. Today, India is alongside the EU, the U.S Japan, Brazil, Argentina, and Switzerland, which have adopted organic standards and put in place an inspection and certification mechanism (K. L. Chadha, 2008)

Nine states in India have promoted policies and programs on organic farming. Uttrakhand has made organic a thrust for improving its mountain agriculture farm economy and livelihood. Mizoram and Sikkim declared their intentions to move to total organic farming. Karnataka has formulated organic policies, and

Maharashtra, Tamil Nadu, and Kerala have supported public–private partnerships for the promotion of organic farming. Indian agriculture market was improved and commercialization based farming has been started by which agricultural market became more competitive and qualitative that is why the organic based product have more customer due to its freshness, nutrition value and microorganism free characteristics. So the organic agriculture expanded gradually (Yadav *et al.*, 2010).

Quality Regulation

Organic agriculture systems and products are required to be certified by accredited agencies to indicate that they have been produced, stored, processed, handled, and marketed in accordance with technical certification adopted. The organic label is a production process claim, as opposed to a product quality claim. Indian has 26 accredited agencies to facilitate the certification of growers. In India, the Tea Board, Coffee Board, Spices Board, and Coconut Development Board have developed guidelines for production and certification, and have encouraged the production and export of organic produce and products.

Objectives of organic farming

1. To produce healthy, nutritious and quality food.
2. To maintain and enhance long-term fertility of soils.
3. To encourage and enhance biological cycles involving microorganisms, soil flora and fauna, plants and animals.
4. To help in soil and water conservation.
5. To minimize all forms of pollution that may result from agricultural practices.
6. To use on farm resources as far as possible.
7. To maintain genetic diversity.
8. To preserve and enhance traditional and indigenous knowledge in farming, varieties.

Characteristics of organic farming systems

Management of Organic farming is focused on the whole farm system and its interactions with climate, environment, social as well as economic conditions, rather than considering the farm as comprises of individual enterprises.

- Protecting the long-term fertility by maintaining organic matter levels, soil biological activity and careful mechanical intervention.

- Nitrogen self-sufficiency through the use of legumes and biological nitrogen fixation, as well as effective recycling of organic materials, including crop residues and livestock wastes.
- Weed, disease and pests control relying primarily on crop rotation, natural predators, crop diversity, organic manuring, use of resistant varieties and limited thermal, biological and chemical intervention.
- Supplementing crop nutrients, where necessary, by using nutrient sources which are made available to the plants indirectly but the action of soil micro-organisms and chemical reactions of the soil.
- The extensive management of livestock, paying full regards to their evolutionary adaptations behavioral needs, and animal welfare issues with respect to nutrition, housing, health, breeding and rearing.
- Careful attention to the impact of the farming system on the wider environment and the conservation of wildlife and natural habitats.

Organic farming system has no bad impact on environment and conserves the natural habitants (Padel and Lampkin, 1994).

Options in organic farming

There are three options going for Organic Farming to alleviate the problems caused by conventional inorganic farming systems. They are is accounts complete exclusion of inorganic fertilizers and pesticides, but advocates.

1. **Pure Organic Farming:** This accounts complete exclusion of inorganic fertilizers and pesticides but advocates the use of organic manures and biological pest control methods.
2. **Integrated green revolution farming**: Under this opinion, the basic trends of the green revolution such as intensive use of external inputs, increased irrigation, development of high yielding and hybrid varieties as well as mechanizations of labour are retained with much greater efficiency on the use of these inputs limited damage to the environment and human health. For this purpose some organic techniques are developed and combined with the high input technology in order to create Integrated Systems such as, "Integrated Nutrient Management" (INM), "Integrated Pest Management" (IPM) and biological control methods which reduce the need for chemicals.
3. **Integrated Farming Systems**: This option involves low input organic farming in which the farmers have to depend on local resources and ecological processes, recycling of agricultural wastes and crop residues.

Hence, for improving the quality of the life and ensuring the reduction in depletion of natural resources needs a farming system, which results in viable and sustainable agriculture production (Maity *et al.*, 2004).

In Indian context, Organic Farming can be more profitable under the following situations, where instead of quantity, quality is more important especially in horticultural crops and vegetables are more important as they are mostly annual in nature.

- Fruits and vegetable crops where use of higher doses of chemical fertilizers (especially N may lead to higher NO3 content) and imbalanced nutrition of crops.
- Plantation crops like tea, coffee, cashew nut etc. where the nutrient removal is less and recycling of these through leaf fall is high.
- Other horticultural crops having high export potential in International markets like spices.
- Local varieties of different crops having high quality and export potentials.
- Neem, dried nuts, oilseeds, pulses, cottons, basmati rice, etc. with export potentials.
- Soils having high fixation capacity of the nutrients like the calcareous, acidic & alkali soils.

Strategies of organic horticulture farming

The future success of organic production would largely depend upon sizeof the farm and supplies of non-chemical inputs, which have to be thoroughly backed up by wellproven package of practices addressing to the objectives of producing vegetable organically. These organic farming practices have to be turn to change in traditional concept of farming. The following issues and their viable strategies are suggested to make organic vegetable production more vibrant, dynamic, and responsive to changing consumer demand both locally and globally as well.

- The research for Organic farming in horticultural crops must be on a system basis. It must be integrated one and must not be looking at in isolation.
- The task of research would be to produce technologies, which can not only increase more food but also more jobs and more incomes.
- The research for organic farming should be focused on developing technologies which may attract the growers to adopt them, keeping in

view of the requirements of small holdings of resource poor small and marginal farmers.

- The research should be in a holistic manner with long-term evaluation of different organic substrates.
- Identification of suitable cover crop and smother crop in a given cropping system.
- There should be strategy for monitoring of changes in groundwater quality with references to heavy metal toxicity, besides nitrate pollution.
- Identification of soil improving crops under major agro-climatic zone.
- Evaluation of soil conservation practices of disease management, change in the habitat for beneficial insects and suitability of trap crops in organic culture and identification of nematode repellant cover crops especially from various horticultural crops should be given due emphasis.
- There is need for marketing research for organically produce for export potential. There should be proper planning for marketing of organically grown fruits, vegetables and food grains that should help farmers to get a better price for their produce. This, in turn, should motivate them to invest more in Organic Farming.
- There should be incentives to the growers who produce organic horticulture products.
- Extension scientists must develop strategy to create interest in small and marginal farmers to adopt organic farming technologies for growing horticultural crops.
- Researchers stated that Organic foods are better in terms of health and safety than the inorganic foods but in the lack of intensive study and research about nutritional and taste characteristics has not been described properly. So the intensive focus on the study of nutritional properties and sensory characteristics of organic food is required (Yadav *et al.*, 2010).

Details of crop production for conversion to organic systems

1. Soil and Water Conservation

Relevant measures should be taken to prevent soil erosion, conservation of water, prevent both excess & improper use of water & pollution of ground as well as surface water. Measures should be taken to prevent salination. Cleaning of land by burning of organic matter should be restricted to the absolute minimum.

2. Choice of Crops & Variety

- Seeds & planting materials should be from traditional / certified organic production.
- Seed treatment should be made with permissible products.
- When certified organic seed not available, chemically untreated seed conventional materials may be used.
- New crop seed & plant material treated with synthetic pesticides, chemicals, related or micro waved can only be allowed in regions where organic agriculture is in the early stage.
- Use of genetically engineered seeds, transgenic plants should not be allowed.

3. Multiple Cropping and Crop Rotation

Mix cropping is the outstanding feature of organic farming in which variety of crops are grown simultaneously or at different time on the same land. In every season care should be taken to maintain legume cropping at least 40%. Mix cropping promotes photosynthesis and avoids the competition for nutrients because different plants draw their nutrients from different depth of soil. The legume fixes atmospheric nitrogen and make available for companion or succeeding crops. Deep rooted plants drew nutrient from deeper layer of soil and bring them to the surface of soil through their leaf fall. So the nutrients leached down to lower strata are further brought back to upper layer by these deep rooted plants. Also help in protecting soil from soil erosion. Farmers should select the crops combination according to their needs and season.

4. Crop Rotation

Crop rotation is the back bone of organic farming practices. To keep the soil healthy and to allow the natural microbial systems working, crop rotation is must. Crop rotation is the succession of different crops cultivated on same land. Follow 3-4 years rotation plan. All high nutrient demanding crops should precede and follow legume dominated crop combination. Rotation of pest host and non pest host crops helps in controlling soil borne diseases and pest. It also helps in controlling weeds. It is better for improving productivity and fertility of soil. Crop rotations help in improving soil structure through different types of root system. Legumes should be used frequently in rotation with cereal and vegetable crops. Green manure crops should also find place in planning rotations. High nutrient demanding crops should always be followed by legume crops and returned back to soil.

5. Use of Biofertilizers and Microbial Cultures

Biofertilizers viz: Rhizobium, Azotobacter, Azospirillum, PSB and Pseudomonas etc have been found to be very effective tools of fertility management and biological nutrient mobilization. Recently customized consortia of such biofertilizer organisms, better adapted to local climatic conditions have also been developed and are available commercially. Efficiency of such microbial formulations is much higher under no-chemical use situations, therefore application of such inputs need to be ensured under all cropping situations.

6. Manurial Policy

Manurial policy should include green manure, leaf litter & vermicomposting. Manure containing human faeces or untreated sewage should not be used on vegetables produced for human consumption.Any organic wastes should be applied in their natural composition & should not be rendered more soluble by chemical treatments. All the materials should be in accordance with the standards. All synthetic nitrogenous fertilizers including urea should be excluded.Organic farm should have manurial policy to include inputs based on microbial, plants or animal origin, provided they do not have adverse effects on the soil and local ecology.

7. Pests, Diseases and Weed Management

Products of traditional nature, preferably prepared at the farm from local plants, animals and microorganisms should be used. Both physical and thermic methods are permitted. Thermic sterilization of soil is allowed to combat both pests & diseases, whenever necessary. All the synthetic herbicides, fungicides, pesticides should be strictly prohibited.

General technology packages for organic horticulture production

1. Timely preparation of soil to a fine tilth with 2-3 ploughings to remove all debris, stubbles, stones, etc. and to avoid infestation of ants and termites. However, minimum tillage is considered as an important component of organic farming.
2. Use of organic manures as basal dose @ 25-38 t/ha through FYM, poultry manures, fish manures, sheep composts etc. Use of organic cakes from neem, groundnut, Pongamia, and castor becomes imperative.
3. Raising of green manure crops like Sesbania or dhanicha and incorporating into the soil, besides using biomass of other plant species.

4. Use of crop residues is essential in organic vegetable, fruits, flowers crops production, which increases the soil organic matter content, maintains soil fertility status, and in turn increases the crop yield.
5. Studies conducted by researchers reported that in Tomato: Application of oil cakes of margosa, castor, and groundnut (@0.2% W/W) is generally is found to reduce the intensity of root gall development. Thamburaj (1994) found that organically grown plants were taller with more number of branches. They yielded 28.18 t/ha, which was at par with the recommended dose of FYM and NPK (120:100:100 kg/ha).
6. Always include legume crop like beans, peas, cowpea, etc in the crop rotation not only to improve the soil fertility by fixing atmospheric nitrogen but else to increase the yield up to 30-35%. Inoculation of legume crop specific rhizobial strains can further improve their N-fixing ability.
7. Choice of varieties should be based on climate and market preference; adopting optimum spacing and timely planting, raising plants/seedling with enough organic manures and bio-fertilizers and using only vigorous seedlings for better establishment, growth and yield.
8. Application of bio-fertilizer is of great significance in organic farming. As they play a nutritional stimulatory and the therapeutic role in improving growth, yield and quality of vegetable crops. Inoculations of vegetable crops with different bio-fertilizers have depicted an encouraging response both in terms of increasing yield, quality and soil fertility. The field response of rhizobium is encouraging as reported by a number of research workers. Azotobacter and Azospirillum depicted a significant influence on vegetable crops, resulting in nitrogen economy of 25-50% and increase in yield from 1-42 %. Similarly phosphorus solubilizers can also save in general 40% phosphorus fertilizers and can enhance the crop yields from 4.7-51%.
9. Use of locally available mulching materials or polythene sheets to reduce moisture loss and minimize weed growth.
10. Use disease resistant varieties that suit the eco-system controlling weeds and removing all infested parts with pests and diseases and raising trap plant to attract insects and to follow crop rotation. For example; cabbage head borer- trap plant as mustard potato/chilli/ tomato rotated with cereals, oilseeds and other vegetables. Control of pests and diseases through use of resistant/tolerant varieties.

11. Use bio-pesticides and bio-control methods for control of various pests and diseases. Natural products like garlic extract are used as broad-spectrum pesticides. Neem, Sabadilla, and Pyrethrum extracts are also used as pesticides. Besides this, bio-fertilizers like Azotobacter, Azospirillwn, PSM, and phosphorus mobilizers have antifungal activities without any residual or toxic effect resulting in the sustainable quality vegetable production. Crop inoculated with mycorrhizal fungi exhibits increased resistance to Rhizoctoniasolani and Fusarium oxysporum. Damping off of tomato caused by Phythium could be considerably prevented. Mycorrhized tomato plants were found more resistant to nematode infection.

Recent advances in organic cultivation

Government is promoting organic farming through various schemes/ programmes under National Mission for Sustainable Agriculture (NMSA)/ Paramapragat Krishi Vikas Yojana (PKVY), RashtriyaKrishiVikasYojana (RKVY), Mission for Integrated Development of Horticulture (MIDH), National Mission on Oilseeds & Oil Palm (NMOOP), Network Project on Organic Farming of ICAR.

In addition to this, Government is implementing a Cluster based programme to encourage the farmer for promoting organic farming called Paramparagat Krishi Vikas Yojana (PKVY) :

- Groups of farmers would be motivated to take up organic farming under ParamparagatKrishiVikasYojana (PKVY). Fifty or more farmers will form a cluster having 50 acre land to take up the organic farming under the scheme.
- In this way during three years 10,000 clusters will be formed covering 5.0 lakh acre area under organic farming. There will be no liability on the farmers for expenditure on certification.
- Every farmer will be provided Rs. 20,000 per acre in three years for seed to harvesting of crops and to transport produce to the market.
- Organic farming will be promoted by using traditional resources and the organic products will be linked with the market.
- It will increase domestic production and certification of organic produce by involving farmers
- In order to implement the Paramparagat Krishi Vikas Yojana in Paramparagat Krishi Vikas Yojana in the year 2015-16, an amount of Rs. 300 crore has been allocated.

Components and pattern of assistance

1. **Adoption of Participatory Guarantee System (PGS) certification through cluster approach**
    - *Mobilization of farmers / local people to form cluster in 50 acre for PGS certification*
    - Conducting of meetings and discussions of farmers in targeted areas to form organic farming cluster @ Rs. 200 / farmer
    - Exposure visit to member of cluster to organic farming fields @ Rs. 200 / farmer
    - Formation of cluster, farmer pledge to PGS and Identification of Lead Resourceful Person (LRP) from cluster
    - Training of cluster members on organic farming (3 trainings @ Rs. 20000 per training)
    - PGS Certification and Quality control
    - Training on PGS Certification in 2 days @ Rs. 200 per LRP
    - Training of Trainers (20) Lead Resource Persons@ Rs. 250 /day/ cluster for 3 days.
    - Online Registration of farmer @ Rs.100 per member cluster x 50
    - Soil sample collection and testing (21 samples/year/cluster) @ Rs. 190 per sample for three years
    - Process documentation of conversion into organic methods, inputs used, cropping pattern followed, organic manures and fertilizer used etc., for PGS certification @ Rs.100 per member x 50
    - Inspection of fields of cluster member @ Rs. 400 /inspection x 3 (3 inspections will be done per cluster per year)
    - Residue analysis of samples in NABL (8 samples per year per cluster) @ Rs. 10, 000/ sample
    - Certification Charges
    - Administrative expenses for certification
2. **Adoption of organic village for manure management and biological nitrogen harvesting through cluster approach**
    - Action plan for Organic Farming for one cluster
    - Conversion of land to organic @ Rs.1000/acre x 50

- Introduction of cropping system; Organic seed procurement or raising organic nursery @ Rs.500/acre/year x 50 acres
- Traditional organic Input Production units like Panchagavya, Beejamruth and Jeevamruth etc. @ Rs.1500 /unit / acre x 50 acre
- Biological Nitrogen Harvest planting (Gliricidia, Sesbania, etc) @ Rs. 2000/acre x 50 acre
- Botanical extracts production units (Neem cake, Neem oil) @ Rs.1000/unit/ acre x 50 acre
- ***Integrated Manure Management***
- Liquid Biofertilizer consortia (Nitrogen fixing / Phosphate Solubilizing/ potassium mobilizing biofertilizer) @ Rs. 500/acre x 50
- Liquid Biopesticides (Trichoderma viridae, Pseudomonas fluorescens, Metarhizium , Beaviouriebassiana, Pacelomyces, verticilium) @ Rs. 500 /acre x 50
- Neem Cake/ Neem Oil @ Rs.500/acre x 50
- Phosphate Rich Organic Manure / Zyme Granules @ Rs. 1000/acre x 50
- Vermicompost (size 7'x3'x1') @ Rs.5000/ unit x 50
- ***Custom Hiring Centre (CHC) charges***
- Agricultural implements (As per SMAM guidelines) - Power tiller, Conoweeder, Paddy thresher, Furrow opener, Sprayer, Rose can, Top Pan balance
- Walk-in tunnels for horticulture (As per guidelines of MIDH)
- Cattle shed / poultry / piggery for animal compost (As per Guidelines of Gokhul Scheme)
- ***Packing, Labeling and Branding of Organic Products of Cluster***
- Packing material with PGS logo + Hologram printing @ Rs. 2500 / acre x 50
- Transportation of organic produce (Four wheeler, 1.5 tone load capacity) @Rs. 120000 max. assistance for 1 cluster
- Organic Fairs (maximum assistance will be given @ 36330 per cluster)

Expected outcomes

The Scheme Envisages

- Promotion of commercial organic production through certified organic farming.
- The produce will be pesticide residue free and will contribute to improve the health of consumer.
- It will raise farmer's income and create potential market for traders.
- It will motivate the farmers for natural resource mobilization for input production.

Source : Department of Agriculture, Cooperation and Farmers Welfare, Ministry of Agriculture and Farmers Welfare, Government of India

References

Anonymous (2016). http://apeda.gov.in/apedawebsite/organic/Organic_Products.htm

Chadha, K. L. (2008). Organic Farming: Concept, Initiatives, Statusand Implication. In Proceedings of the Padmanav Panda Memorial Orationat Orissa University of Agriculture & Technology,Bhubaneswar,India,25 July 2008.

Dale Rhoads, ''Purdue Extension Service'', What is organic farming?

Danielle Treadwell, Jim Riddle, Mary Barbercheck, Deborah Cavanaugh-Grant, Ed Zaborski, ''Cooperative Extension System'', What is organic farming?

Diver, Steve. (2014). "Controlled Microbial Composting and Humus Management: Luebke Compost". Retrieved 11 May 2014.

FAO (1999a). "Organic agriculture." COAG/99/9 Rome: FAO.

Funtilana, S. (1990). Safe, inexpensive, profitable, and sensible. International Agricultural Development, March-April 24.

H. Martin, ''Ontario Ministry of Agriculture, Food and Rural Affairs'' Introduction to Organic Farming, ISSN 1198-712X

IFOAM(International Federation of Organic AgricultureMovements) (1997). The Future Agenda for Organic Trade: Proceedings of the 5th IFOAM International Conference on Trade in Organic Products.Tholey-Theley, Germany: IFOAM.

Lampkin, N. (1990). Organic Farming. Farming Press Books, Ipswich, UK.

Maity T K and Tripathy P. (2004). Organic farming of vegetables in India: problems and prospects, Department of Vegetable Crops, Faculty of Horticulture, Bidhan Chandra KrishiViswavidyalaya (WB). (www.share4dev.info/kb/documents/2997.pdf). Access date: 28 Sept, 2013.

Nayler, Justin. (2014)"Second Thoughts About Organic Agriculture" (PDF). Soil And Health Library. Retrieved 11 May 2014.

Padel S and Lampkin N. (1994). Conversion to organic farming: An overview: The economics of organic farming. An international perspective: CAB International, Wallingford, pp. 295-310.

Paull, John (2011). "Attending the First Organic Agriculture Course: Rudolf Steiner's Agriculture Course at Koberwitz, 1924". *European Journal of Social Sciences'*. 21 (1): 64–70.

Paull, John (2011). "The secrets of Koberwitz: the diffusion of Rudolf Steiner's agriculture course and the founding of biodynamic agriculture". *Journal of Social Research & Policy*. 2 (1): 19–29.

Paull, John (2013). "Breslau (Wroc³aw): In the footsteps of Rudolf Steiner", Journal of Bio-Dynamics Tasmania, 110:10-15.

Paull, John (2013). "Koberwitz (Kobierzyce); In the footseps of Rudolf Steiner", Journal of Bio-Dynamics Tasmania, 109 (Autumn), pp. 7-11.

Paull, John (2010). "From France to the World: The International Federation of Organic Agriculture Movements (IFOAM)", *Journal of Social Research & Policy*, 1(2):93-102.

Sikkim Becomes India's First Organic State. (2016). Available online: http://www.thehindu.com/news/national/sikkim-becomes-indias-ûrstorganic state/article8107170.ece (accessed on 14 January 2016).

USDA. (1980). Report and Recommendations on Organic Farming. USDA, Washington, D.C.

Yadav AK. (2010). Organic Agriculture, Concept, Scenario, Principals and Practices. Director National Centre of Organic Farming, Ghaziabad National Centre of Organic Farming Department of Agriculture and Cooperation, Ministry of Agriculture, Govt of India, Uttar Pradesh.

Yadav, A.K. (2015). OrganicAgricultureataGlance. Availableonline:http://krishijagran.com/farm/scenarioin-india/2015/03/Organic-Agriculture-At-a-Glance (accessed on 2 April 2015).

2

Role of Indigenous Technical Knowledge in Organic Fruit Production

M. Surendra Babu T. Thomson, G. Koteswara Rao and M.M. Nagaraju

Department of Plantation crops and Spices, College of Horticulture, Vellanikkara Kerala Agricultural University, Thrissur-680656, Kerala

Abstract

Organic farming is a method of crop and livestock production that involves much more than choosing not to use pesticides, fertilizers, genetically modified organisms, antibiotics and growth hormones. Organic fruit production essentially excludes the use of many inputs associated with modern farming, most notably synthetic pesticides and fertilizers. Organic agriculture, the innovative farming system can build-on and enhance the traditional knowledge and practices of local and indigenous communities. Indigenous Knowledge can be broadly defined as the knowledge that anindigenous (local) community accumulates over generations of living in a particular environment. ITK in fruit production is important for reducing the cost of cultivation, reducing the pollution and also the natural resource base. It is very important to understand these age old practices to address the local problems and to document it before is vanished away.

Keywords : *Organic farming, ITK, fruit production*

Introduction

- Organic fruit production essentially excludes the use of many inputs associated with modern farming, most notably synthetic pesticides and fertilizers.

- To the maximum extent possible, organic farming systems rely upon crop rotations, crop residues, animal manures, legumes, green manures, off-farm organic wastes, mechanical cultivation, mineral- bearing rock powders and biological pest control.
- These components maintain soil productivity and tilth, supply plant nutrients and help to control insects, weeds and other pests.
- India, the country of *Rishi* and *Krishi,* has a very rich heritage of traditional health control methods and several treatment systems (Ayurvedic, Unani, Homeopathy).
- Moreover, history of Indian agriculture shows that most of its farming community relies on traditional methods and ecological agriculture such as *'agnihotra'* and *'panchagavya'* the farming systems based on ancient techniques for soil and animal management.
- Organic agriculture, the innovative farming system can build-on and enhance the traditional knowledge and practices of local and indigenous communities.
- This means farmer's knowledge of ecological systems, environment and their conventional wisdom has to play more role in making organic farming more sustainable.

What is Indigenous Technical Knowledge

- Indigenous Knowledge can be broadly defined as the knowledge that anindigenous (local) community accumulates over generations of living in a particular environment.
- Indigenous Technical Knowledge : Is specifically concerned with actual application of the thinking of the local people in various operations of agriculture and allied areas.
- Indigenous Technical Knowledge (ITK) is the actual knowledge of a given population that reflects the experiences based on tradition and includes more recent experiences with modern Technologies.
- ITK is the information base for a society, which facilitates communication and decision making.
- ITK are dynamic, and are continually influenced by internal creativity and experimentation as well as by contact with external systems.
- Indigenous technical knowledge (ITK) that has the element of use of natural products to solve the problems pertaining to agriculture and allied activities.

- ITK is the information base for a society, which facilitates communication and decision making.
- This resource was passed from generation to generation through orally, cultural traditions like festivals, foods habits etc.
- Through this, farmers were conserving the natural resources and they were managing soil health, pest and disease effectively.
- Examples: - Burning crackers during the evening of Deepavali was mainly started to control field pests during cropping season - Broadcasting foods materials during SARGA (A north Karnataka agriculture festival) is to attract birds in the field, which in turn will catch the field pests.
- Mixed cropping is one of the best practices followed to manage nutrients in the soil and also to control pest population.
- Hence it is very important to understand these age old practices to address the local problems and to document it before is vanished away .

ITK for Weather Prediction

Weather prediction through birds and other animals

Farmers also predict weather by observing closely the different activities of various birds, animals etc. The following are some indigenous beliefs :

- It is believed that on a hot summer day the cry of the bird called "Nialu" for water brings rainfall.
- During the rainy season farmers observe the "Matilari" bird (House swift) and they predict heavy rainfall if the bird flies high in the sky.
- If the Maina bird bathes in the water it indicates that there will be rainfall within one or two days.
- During long hot days in summer if the cry of theapiha bird is heard then people believe that God will quench her thirst and there will be rainfall after one or two days.
- A group of sparrows frolicking in the sand indicates that there will be rainfall that day or the next day and if they are observed to be playing in water then it is believed that the weather will be dry for some days to come.

ITK Practices in Organic Fruit Production

- Panchagavya, Dasakavya, Jivamrute, Dashparni Ark, Amrutpani

Panchagavya

In Sanskrit, Panchagavya means the blend of five products obtained from cow. Panchagavya is made from five products of the cow — its dung, urine, milk, ghee and curd 20 litres of panchagavya you need the following :

1. Fresh cow dung -5kg
2. Fresh cow's urine–3 litres
3. Cow's milk–2 litres
4. Cow's curd – 2 litres
5. Cow's ghee – 500 gms
6. Jaggery – 500 gms
7. Water or sugarcane juice-3 litres
8. Ripe banana fruit -1 bunch (12 nos.)
9. Tender coconut water–3 litres

(You can ferment 3 litres of tender coconut water by keeping it in a pot for 1 week.)

Method of preparation

- Take 5 kgs of fresh cow dung and mix it thoroughly with 500 gms of cow's ghee and keep it in a plastic drum or a mud pot. It should not be kept in a metal container because it will corrode and react with the metal. Keep the ghee and the cow dung mixture for 4 days, mixing it twice a day.
- On the 5th day add cow's urine, cow's milk (cow's milk can be boiled, cooled and then added) and cow's curd, then jaggery with water or sugarcane juice and banana fruit (which has to be mashed and mixed thoroughly.) Then tender coconut water has to be added.
- Wait for another 15 days, stirring twice daily. Stirring the contents for about 20 minutes each time facilitates aerobic microbial activity.
- On the 19th day the panchagavya solution will be ready. This solution must be kept under a net, i.e. it must be covered with a muslin or fine cloth so that the common fly cannot sit on it and lay eggs.
- For use after 19th day, stir the mixture at least once a day to aerate it.

Periodicity of use

- Once in 15 days, it can be used for all crops. In winter crops, 1.5-2% usage is sufficient but for all other crops 3% should be used.
- 3% solution was found to be most effective compared to the higher and lower concentrations investigated. Three litres of Panchagavya to every 100 litres of water is ideal for all crops.
- Growth promotion is favoured by the beneficial micro organisms multiplied in large quantities after application.

Effect of Panchagavya

1. Leaf

Plants sprayed with Panchagavya invariably produce bigger leaves and develop denser canopy. The photosynthetic system is activated for enhanced biological efficiency, enabling synthesis of maximum metabolites and photosynthates.

2. Stem

The trunk produces side shoots, which are sturdy and capable of carrying maximum fruits to maturity. Branching is comparatively high.

3. Roots

The rooting is profuse and dense. Further they remain fresh for a long time. The roots spread and grow into deeper layers were also observed. All such roots help maximum intake of nutrients and water.

Beneficial effects of panchagavya on commercial fruit crops

1. Mango

- Induces dense flowering with more female flowers
- Irregular or alternate bearing habit is not experienced and continues to fruit regularly
- Flavour and aroma are extraordinary

2. Acid lime

- Continuous flowering is ensured round the year
- Fruits are plumpy with strong aroma

3. Banana

- Spraying Panchakavya 3%To get high yield without external chemical inputs

- In addition to adding with irrigation water and spraying, 3% solution (100 ml) was tied up at the naval end of the bunch after the male bud is removed. The bunch size becomes uniform. One month earlier harvest was witnessed. The size of the top and bottom hands was uniformly big.

Application of waste cotton from ginning factory reduce the weed growth. And also Withstand water

2. Dasakavya

- Dasagavya,: is an organic preparation made from ten products in the form of panchagavya and certain plant extracts
- The plant extracts are prepared by separately soaking the foliage in cow urine in 1:1 ratio (1 kg chopped leaves in 1 litre cow urine) for ten days.
- The filtered extracts of all the plants are then added @ 1 litre each to 5 litre of the panchagavya solution.
- The mixture is kept for 25 days and stirred well, meanwhile, to ensure thorough mixing of panchagavya and the plant extracts.

Mode of use

The Dasagavya solution is filtered to avoid clogging of sprayer nozzles and is recommended as foliar spray at 3 % concentration. Soaking of seeds or dipping the roots of seedlings in 3 % solution of dasagavya for 20 minutes before planting enhances seed germination and root development.

Advantages

- Increases growth, yield and quality of the fruit crops.
- Controls pests like aphids, thrips, mites and other sucking pests.
- Controls diseases like leaf spot, leaf blight, powdery mildew etc. Spraying 3 % dasagavya in rose controls thrips and powdery mildew in grape.

Jivamrute

The Jivamrute contains

1.	Cow dung	–	60 kg
2.	Cow urine	–	10 lit.
3.	Black juggary	–	2 kg
4.	Gram floor	–	2 kg
5.	Water	–	200 lit.

This mixture keeps for fermentation for 7 days.

After that it mixed with irrigation water which will sufficient for one acre. After applying this there is no need to apply any other chemical fertilizer and yield is same that obtained from the application of chemical fertilizer.

4. Dashparni Ark

- Used for control of all type of pest
- Take 100 liter of water
- Add fresh leaves' crush of following plants

	Plant Name	Quantity (Kg)
1)	KaduNimb (Azadtractaindica)-	5
2)	Papaya	2
3)	Karanj	2
4)	Sitafal	2
5)	Jatropha	2
6)	Kanher	2
7)	Dhatura (White)	2
8)	Rui (Calotrophis)	2
9)	Nirgudi / Gulwel	2
10)	Ghaneri (*Lantena camera*)	2

- Add 5kg cowdung& urine each
- Add 2 kg hot green chilli crush & 250gm garlic crush
- Incubate for 30 days with daily stirring
- Use @250 ml in 10 lit. of water for spraying

Amrutpani

Preparation of Amrutpani

200 litres of, Amrutpani is needed for one acre of farm land, which is prepared as under :

Materials required

1. Quarter kilo of ghee (clarified butter) from a desi (indigenous breed) cow.
2. Half kilo of honey

3. 10 kilo of fresh dung from a desi cow
4. 200 litres of water

How to use Amrutpani?

Thoroughly mix quarter kilo of ghee into 10 kilos of cow dung. Blend half kilo of honey into this mixture and add 200 litres of water stirring all the time. The mixture thus obtained is Amrutpani crops where the seedlings are transplanted, the roots can be dipped into Amrutpani before planting.

ITK Practices Crop-Wise

1. Grape

- Groundnut cake is powdered, soaked in water and poured to grapes @ 1 bucket/pit for better fruit quality and yield.
- Circular trenches are dug around vines in which green manures and FYM are applied in 3:1 ratio and covered with soil before the monsoon starts in order to get higher yields.
- Neem cake is applied to control the nematodes.
- 40 kg of neem cake per acre is applied as basal manure for grape crops to prevent diseases

2. Mandarin Orange

- Orange trees attacked by Stem borer are given time wash, holes are cleaned and plugged with lime soaked cotton or wrapped with lemon grass.
- Greenish *Aloe vera* plants are cut into small pieces and spread to a radius of 2 feet around the tree during flowering to control powdery mildew.
- Collected orange seeds are mixed with ash to avoid ants' attack

3. Guava

- ITK Practiced Pounding 2kg.of foliage of *Calotropis sp.* with 3 kg.of neem cake, soaking them in 20 lit. of water for four days and dissolving the extract in 200lit.of water and spraying for one ac. to control all the pests.
- Apply pig manure @ 1 basket/tree to increase the yield and to prevent flower shedding.
- Dried neem fruits are powdered and applied @ 500g. per tree to control the nematode attack.

4. Banana

- To control fruit rot in banana during storage, the fruit stalks are soaked in 10% thulasi*(Ocimumcanum)* leaf extract or 1 % neem oil solution and stored.
- For quick ripening of banana fruits, lime solution is sprinkled over the bunches.
- Banana suckers are immersed for a while in 1 lit. ofneem oil dissolved in 100 lit. of water before planting in order to prevent rhizome rot.

5. Mango

- Neem oil is sprayed to control the hoppers.
- Sunflower is cultivated in between the mango trees to attract honey bees which increase pollination and fruit production.
- To induce early ripening of mango fruits, they are spread on a layer of the branches of 'Aavaram' *(Cassia auriculata)* plant on the floor and again covered with its branches and finally

ITK Operational Wise

1) Soil and Water Management Techniques
2) Outer shells of tamarind fruits are applied in the field to control Cyperusrotundus.
3) About 10kg. of dried cow dung is ground into fine powder and mixed with ash dusted in the early morning, to control pests and diseases.
4) Calotropis leaf extract is applied at the place of termite attack to control them.
5) Planting vettiver*(Vetiveriazizanoides)* slips across the slope or around the fields to prevent soil erosion.

Preparatory Cultivation

1. Achieving fine tilth is better than applying manures.
2. It is better to plough intensively than extensively.
3. It is better to have deep ploughing rather than shallow ploughing for good crop growth.
4. Plough four times for garden land and seven times for wetland.

Seed and Sowing

1. Yield of almost all the crops depends on seed quality.
2. Even if it is a quality seed, it should be a dried one because well dried seeds will have higher longevity and keeping quality.
3. To maintain the seed viability and prevent it from outside damalt is better to start planting from 'Sanimoolai' (north east) of the field to get higher yields.
4. It is better to perform sowing and planting operations during evening hours.
5. Even if the seeds are of good quality, they must be sown in the right season only.
6. Sowing in the right season is better even on a poor land.

Manures

1. Organic manures always encourage good crop growth.
2. Green manure crops like daincha ,sun hemp etc., are raised and ploughed insitu before their flowering.
3. Red gram is also used as a green manure crop *which* improves the soil fertility.
4. Sheep penning results in more crop yields.
5. *Goat* manure gives benefits to crops grown in the same season.
6. Poultry manure serves as a good source of crop nutrients.
7. Cow urine is more nutritious than cow dung.

Weed Management

1. Cultivating rice once in three years in garden lands to control 'Arugu' *(Cyanodondactylon)* weeds.
2. 'Raising and ploughing the green manure crops tike daincha..
3. Growing horse gram to control nut grass *(Cyperusrotundus).*
4. Growing cowpea as a green manure to control nut grass.

Pest and Disease Management

1. Crops grown in alkaline soils are more prone to disease attack.
2. Kitchen ash is applied to control aphids.
3. Mixing cow urine, neem oil and tobacco decoction together and spraying on crops controls sucking pests.

Storage Pest Control

1. Grains, trees etc. harvested on full moon day are prone to storage pest attack.
2. Mixing the dried leaves of neem with seeds while storing them.
3. Mixing the dried leaves of notchi (*Vitexnegundo)* with seeds while storing them.
4. Storing the seeds after mixing with pungam (*Pongamiapinnata*) leaves.

Conclusion

ITK in fruit production is important for reducing the cost of cultivation, reducing the pollution and also the natural resource base.

ITK is the basis for self sufficiency and self determination for at least two reasons *i.e.*

1. farmers are familiar with indigenous practices and technologies. They can understand, handle and maintain them better than introduced practices and technologies .
2. ITKs are drawn on local resources, farmers are less dependent on outside supplies which can be costly, scarce and available only irregularly.

References

Anonymous (2006) http://en.wikipedia.org/wiki/Panchgavya.

Anonymous (2011). www.uttanKrishisanshodhansansthan org.

Gopal Kumar Niroula Chhetry and Lassaad Belbahri (2009). Indigenous pest and disease management practices in traditional farming systems in north east India. *Journal of Plant Breeding and Crop Science* Vol. 1(3). pp. 028-038.

Selvaraj, N., B. Anitha, B. Anusha and M. Guru Saraswathi (2007). Organic Horticulture. Horticultural Research Station, Tamil Nadu Agricultural University, Udhagamandalam-643 001.

Subrahmanyeswari, B and Chander, M. (2013). Integrating indigenous knowledge of farmers sustainable organic farming. *Indian journal of Traditional Knowledge* 12(2):259-264.

3

Organic Production of Horticultural Crops in North East Region of India

L.C. De

ICAR-NRC for Orchids, Pakyong-737106, Sikkim, India

Introduction

India is bestowed with wide range of climate and physic-geographical conditions and as such is most suitable for growing various kinds of horticultural crops such as fruits, vegetables, flowers, nuts, spices and plantation crops (coconut, cashewnut and cocoa). Its horticulture production has increased significantly over the last two decades and as per the final estimates, by 2013-14, it has increased to about three times (2.87) since 1991-92 and to about twice (1.90) compared to the production in 2001-02. This has placed India among the foremost countries in horticulture production, just next to China. As per National Horticulture Database 2014, during 2013-14, India's contribution in the world production of fruits & vegetables was 13.6% & 14%, respectively. Total production of fruits during 2013-14 was about 89 million tonnes while that of vegetables was 163 million tonnes.

India is the largest producer, consumer and exporter of spices. It is the second largest producer of fruits and vegetables in the world and occupies first position in the production of fruits like mango, banana, papaya, sapota, pomegranade, acid lime and aonla and vegetables like peas and okra. It is next only to china in production of many vegetables like potato, tomato, onion, brinjal, cabbage, cauliflower and broccoli etc.

Besides meeting the increasing demand of the domestic population, which continues to grow, India exports some portion of its horticulture produce. During 2013-14 total exports of horticulture produce by India was 3.69 million metric tonnes which amounted to about Rs 143.6 billion. Even though the quantum of export decreased in comparison to the year before i.e. 2012-13 when it was 3.7 million metric tonnes , the value of export of horticulture produce increased by about 35.6% from Rs 105.9 billion in 2012-13. Except for 2010-11 when the exports of horticulture declined by about 7%, the export of horticulture produce has seen an increased during the last six years.

Agriculture scenario in North East Region

The economy of the North Eastern states is mainly rural and agrarian. The region offers scope for cultivation of a wide variety of agricultural crops because of its diversities in topography, altitude and climatic conditions. The extent of cultivable land in the NE region varies from state to state. Land is a critical resource in many of the NE states, and availability and management of land for agricultural activities are essential for raising the region's overall agricultural production and productivity. The region's agricultural system is predominantly traditional. The overall geographical land to man ratio for the NE region (0.67 hectare/person) is much higher than the national average (0.32 hectare/person). Population to land ratio is highest highest in Arunachal Pradesh followed by Mizoram, Sikkim and Manipur.

The per cent utilization of cultivable area in the NE regions (62.04%) less than the national average (73.05%). About 80% of the farmers in the NE region belong to small (less than 1.44ha) and marginal (less than 0.40 ha) category. Moreover, with increase in population, the average size of land holding is gradually reducing over the years. This is primarily because hilly terrain constitutes nearly two third of the regions geographical area, and large sized holding are not feasible. The average size of land holding for the NE States (1.60 ha) is marginally higher than the all India (1.57 ha). Among the NE States the average size of land holding is highest in Nagaland (6.92 ha) and lowest in Tripura (0.97ha).The average plot size is very small for mechanization of agriculture and adoption of modern farming practices. Hence, high value crops such as different types of flowers will provide high remuneration from limited resources.

Horticulture Scenario in North East Region

There is immense potential for vertical and horizontal growth in horticulture sector in the region. At present horticultural crops account for only 18.60% of cultivated area (Table 1). This share is highest in Sikkim followed by Manipur, Arunachal Pradesh, Meghalaya, Tripura, Mizoram, Assam and Nagaland. There is need to expand area under horticultural crops particularly in Assam, Mizoram and Nagaland where at present it is less than 20% of the cultivated area.

In terms of its contribution to the national production, the Region accounts for about 5.1% (fruits) and 4.5% for vegetables.

Table 1: Percent Area under Horticulture Production

State/UT's	% Area under Horticulture
Arunachal Pradesh	36.26
Assam	15.49
Manipur	37.05
Meghalaya	30.44
Mizoram	16.99
Nagaland	4.66
Sikkim	50.92
Tripura	27.52
Total	18.66

Source : Directorate of Economics and Statistics & National Horticulture Board

In the NEH Region, the total area under horticultural crops is around 857.8 thousand hectare which is around 3.14 per cent of the total geographical area of the region and it gives total production of 7815.9 thousand tonnes.

Crop wise area, production and productivity of different horticultural crops of NE region (2010-11) is given below:

Crops	Area (ha)	Production (MT)	Productivity (MT/ha)
Fruits	344,379	2950,328	8.57
Vegetables	444,603	4502,091	10.13
Spices	68,858	363,527	5.28

Source : NHB and State Departments (2010-11)

Within the horticulture sector in the NE Region, fruit crops occupy 40.14%, vegetables 51.83% and spices 8.03% area. From the production point of view, fruit crops contributed 37.75%, vegetables 57.60% and spices 4.65% to the horticulture production basket of the NE Region. Considering the excellent climatic conditions, abundant rainfall and fertile soil (high organic content) of the region, the productivity of different horticultural crops is quite low as compared to national productivity.

Among fruits, citrus has the largest area under cultivation (34%), followed by banana (24%), pineapple (17%), apple (4%), papaya, mango, litchi, passion fruit (3% each), guava (2%) andkiwi (1%). Other fruits have 6% of the total horticulture area under cultivation. In terms of production, banana has the largest share (34%), followed by pineapple (24%), citrus (21%), papaya (6%), mango, guava, passion fruit (3% each), litchi (2%), grapes and apple (1% each).

Among vegetables, potato has the maximum area (27%) under cultivation followed by cabbage (10%), peas (7%), cauliflower (6%), tomato and brinjal

(5%), okra and chilli (4%), onion (2%) while other vegetables have 30% of the total area under vegetable cultivation.

In terms of production, potato has the highest production and contributes 22% to vegetable production in NER followed by other vegetables (22%), cabbage (18%), tomato (11%), cauliflower (9%), brinjal (7%), okra (4%), peas and chilli (2% each) and onion (1%).

The major horticultural crops of North East Region is given in Table 3.

Statewise area, production and productivity of horticultural crops of NE Region (2010-11) is listed Table 2.

Table 2 : State wise, production and productivity of horticultural crops in North East Region (2010-11)

Crops	Area (ha)	Production (MT)	Productivity (MT/ha)
Arunachal Pradesh	101,967	417,849	4.10
Assam	401,009	4,545,288	11.33
Manipur	78,349	671,854	8.58
Meghalaya	73,076	552,104	7.56
Mizoram	69,786	490,936	7.03
Nagaland	27,473	251,145	9.14
Sikkim	44,375	188,290	4.24
Tripura	66,953	802,211	11.98

Source : NHB and State Departments

Table 3: Major Horticultural crops in North East Region

States	Horticultural crops
Assam	Banana, Citrus, Guava, Pineapple, Papaya, Jackfruit, Arecanut, Potato, Guava, Litchi, Cabbage, Cauliflower, Brinjal, Pea, Tomato, Ginger, Turmeric
Nagaland	Citrus, Pineapple, Banana, Papaya, Potato, Tomato, Cabbage, Okra, Ginger
Tripura	Citrus, Banana, Litchi, Pineapple, Peach, Papaya, Brinjal, Potato, Cauliflower, Tomato, Okra
Meghalaya	Pineapple, Citrus, Banana, Papaya, Guava,, Passion fruits, Strawberry, Kiwi, Peach, Potato, Tomato, Brinjal, Cabbage, Cauliflower, Ginger, Turmeric
Arunachal Pradesh	Apple, Citrus, Banana, Pineapple, Guava, Kiwi, Peach, Passion fruits, Large cardamom, Black pepper, Ginger, Potato, Tomato, Cabbage
Manipur	Pineapple, Papaya, Banana, Jackfruit, Citrus, Peach, Kiwi, Passion fruit, Cabbgae, Pea, Tomato, Cauliflower, Chilli, Ginger, Turmeric
Mizoram	Citrus, Banana, Passion fruit, Grape, Peach, Papaya, Chayote, Bitter gourd, Bean, Brinjal, Ginger, Turmeric, Birds Eye Chilli
Sikkim	Citrus, Kiwi, Banana, Potato, Tomato, Cabbage, Broccoli, Cauliflower, Pea, Okra

Organized cultivation of crops like kiwi, passion fruit, off season vegetables, anthurium, cut flowers (Rose), patchouli, geranium etc. has stared in recent years. While food grains are grown in the valleys (plain and gentle slopes), horticulture crops are cultivated on higher hill slopes. Considerable diversity exists among the regional horticultural species including variation in plant type, morphological and physiological characteristics, reactions to diseases and pest adaptability and distribution. In addition to their nutrition value, many regional horticultural crops are also used for medicinal purposes. The diversity of major crops in NER is recorded as :

Taro - 300 species

Yams - 200 species

Citrus - 17 species

Banana - 16 species and 120 variants

Medicinal Aromatic Plants and Herbs - 306 species

The annual linear rates of growth in area and production of horticultural corps in the regionwork out as below, with considerable variations across the States:

Annual Growth Rate Per cent

	Area	Production
Fruits	9.7	11.2
Vegetables	8.8	14.8

Lesser known fruit crops of NEH Region

Meghalaya: *Prunus nepalensis, Citrus indica, Myricanagi, Eleagnuskhasianum, Flemingia vestits, Docynia indica khasiana, Citrus macroptera, Citrus latipes.*

Assam: *Citrus lemon, C. jambhiri, C. megaloxycarpa, C. assmensis, Artocarpus lakoocha, Dillenia indica, Averhoea carambola.*

Nagaland: *Citrus ichangensis, C. aurantium, Musa magnesium, Malusbaccata.*

Sikkim: *Musa sikkimensis, Baccaurea sapida, Docynia indica*

Tripura: *Psidium guinense, Zizyphus mauritiana, Citrus megaloxycarpa, Phyllanthus acidus, Flacourtica indica, Baccaurea sapida, Averrhoea carambola*

Arunachal Pradesh: *Garicinia cowa, Baccaurea sapida, Citrus medica, Sterculia hamiltonii, Actinidia chinensis, Mangifera sylvatica, M. chassina, Musa ornata, Musa velutina, Nepheleum lappaceum, Lithocarpus* spp., *Castanopsisindica, Fragaria* spp.

Wild edible vegetable of NEH region

Chenopodium spp., *Rhynchotechumelliptium*, *Houttuyniacordata*, *Clerodendrumcolebroekianum*, *Erygiumfoetidum*, *Parkiaroxburghii*, *Momordica cochinchinensis*, *Momordica dioica*, *Sechium edule*, *Vigna vexillata*, *Solanum depressum*, *Solanum sepentinum*, *Psophocarpus tetragonolobus*, *Allium hookerii.*

High potential medicinal plants of NE region

Panax psuedoxinseng - AIDS

Hydnocarpus kurzii – Leprosy

Litseacubeba – Paralysis

Clerodendrum colebrookianum– Heart disease

Coptisteeta – Malaria

Vitextrifolia – Tuberculosis

Aconitum heterophyllum – Diabetes and rheumatism

Alpinia galanga – Skin disease

Cucurma caesia – Swellings, sprains

Taxus baccata – Breast cancer

Acorus calamus – Influenza, headache, cough, cold

Ambrosia artimisifolia – Wounds, cuts

Antidesma brunius– Syphilic ulcers

Achyranthes aspera – Leprosy

Horticultural crops used in NTFP

1. Fibre: Okra, pineapple, snake plant, banana
2. Fodder: *Artocarpus lakoocha*, *Prunus cerasoides*, *Ficushookerii*
3. Dyes: *Cucurma*, *Myricanagi*, *Garcinia cowa*
4. Essential oils: *Alpinia, Cinnamomum, Cymbopogon, Ocimum, Pogostemon,* Vanilla
5. Phytochemicals: Olive, Citrus, Coconut, Pineapple, Papaya, Jackfruit, Apple.

Technological Interventions

Concerted research efforts have been made by research institute to identify a large number of improved varieties and production technologies of fruits, vegetables, tubercrops including potato and plantation crops suitable for the region.

Fruit crops

Based on the survey conducted in Meghalaya, Arunachal Pradesh, Mizoram, Sikkim and Assam, a large number of citrus species were collected and analyzed for physiochemical characteristics. Suitable rootstocks for grafting and budding of different types of citrus have been identified. Varieties of different fruit crops like guava, peach, passion fruit, kiwi, strawberry, banana, pineapple, litchi, mango, papaya etc. have been identified.

Citrus : Khasi mandarin, Assam lemon, Acid lime, Sweet orange.

Peach : TA 170, Floridasun, Shan-e-Punjab, Sharbati.

Plum : Santarosa, Alton, Titrok

Pear : Bartlett, Baghughosa, Sand Pear, Patharnakh Red

Apple : Royal Delicious, Rich-a-Red, Red Delicious, Golden Delicious

Kiwi : Bruno, Allison, Abbot, and Hayward

Guava : Allahabad Safeda, L-49, and H-7.

Strawberry : Sweet Charlie, Ofra, and Chandler

Passion fruit : Purple, Yellow, Kaveri

Banana : Jahaji, Borjahaji, Chinichampa, Malbhog.

Litchi : Sahi, China, Bedana and Rose Scented

Pineapple : Kew, Queen.

Mango : Amrapalli.

Papaya : Honey Dew, CO-2.

Propagation techniques in guava, peach, citrus, passion fruit and kiwi have been standardized (Table 4).

Table 4 : Standardization of propagation technique of major fruit crop in the region

Crop	Rootstock	Standardized technique
Citrus	C.latipes, C. limonia, C. volkameriana, C.grandis	*Softwood grafting, wedge grafting*
	C. medica, c. latipes, C.jambheri, C. volkamariana	*Shield budding*
Peach	Local peach	Tongue and Wedge grafting
Guava	Psidiumguineense	*Patch budding*
Kiwi		Cuttings
Passion fruit		Cuttings

High density plantation in banana, pineapple and guava have been standardized. In banana, cv. 'Dwarf Cavendish' is found effective in the spacing of 1.5m x 1.5m under Arunachal Pradesh climate. In pineapple, double row system of planting across the slope at the spacing of 30 x 45 x 90 cm is found to accommodate 45, 000 plants/ ha and to produce 40-50 tonnes yield per hectare.

An integrated approach in rejuvenation of citrus especially in Khasi mandarin and acid lime has been developed through preparation of half moon terraces and cleaning of basin, adoption of sanitation processes like removal of water shoots, dry and diseased twigs, lichens, mosses and parasites and orchard management practices like nutrient management practices schedule, IPM, moisture conservation measures and dehorning at 1 m and 1.5 m height.

Vegetables

In vegetable crops, improved varieties and hybrids have been identified suitable for varied agroclimates (Table 5). Package of practices for tomato, capsicum, French bean, brinjal, cabbage, cauliflower, sweet potato, colocasia, turmeric, ginger and chilli have been developed. Organic cultivation of ginger, turmeric and cole crops has been initiated. Some F_1 hybrids of brinjal and tomato, resistant to bacterial wilt are also in the advance stage at ICAR research Complex, Umiam.

Table 5: Identified varieties of vegetable crops for NEH Region

Crop	Varieties / Hybrids
Potato	KufriKhasiGaro, KufriJyoti, KufriGiriraj, KufriMegha
Brinjal	Bholanath, Singhnath, PPC, Sel-5, RCMBL –1
Capsicum	Indame-3, Mahabharat, KT-1
Tomato	ArkaAlok, ArkaAbha, Manikhamnu, Manileima, Manithoibi, Rupali, Avinash – 2, Pusa Hybrid – 2, Rocky, cherry tomato.
Cauliflower	Snowball – 16, PusaShubhra, Himani, Pusa Himjyoti, Pusa Synthetic.
Cabbage	Pusa Mukta, Pusa Ageti, Green Express, Pride of India, Challenger
Broccoli	Fiesta, KTS-1, Pushpa,

Contd.

Radish	JapaneseWhite, Chinese Pink, Meghalaya Local
Okra	7-Dhari, ParbhaniKranti, ArkaAnamika
Carrot	Nantes, Pusayamdagni, Pusa Meghali, Hybrid-1
Chilli	Agni, Bird's Eye Chilli, DulleKhursani, Gauhati Black, Raja Mircha, TejpurChilli
French Bean	RCMFB – 1, ArkaKomal, PusaParvati
Cucumber	Poinsette, Pusa Sanjog
Sweet Potato	Sonipat –2, SreeBhadra, S-162
Colocasia	ML-1, ML-9
Pea	Arkel, Lincoln, PM-2, VL-3

Spices

Tumeric: Lakadong, Megha Turmeric – 1

Ginger: Nadia, China, Varada

Large cardamom: Ramla, Red Sawney, Belak, Golsey, Madhusey, Ramsey.

In plantation crops coconut-arecanut based cropping system involving spices and fruit crops have been developed for the region.

Arecanut: Mangala, Sumangala

Black pepper: Panniyur, Hybrid-1-7, Panchami, Pournami

Floriculture

The region is rich in ornamental palnts like orchids, *Bauhinia, Cassia, callistemon, Erythrina, Jacaranda, Magnolia, Rhododendron, Azalia, Bougainvillea, Camelia, Gardenia hibiscus, Jatropa, Nerium, Thumbergia.* Out of 1350 orchid species reported, 900 species occur in NE region. Among orchid species, *Vanda teres, Vanda coerulea, Renanthera imschootiana, Paphiopedilum, Dendrobium, Cymbidium, Cattleyas, Rhyncostylis retusa* are common. This region is potential for production of commercial flowers like orchids in Meghalaya, Arunachal Pradesh and Sikkim; anthurium in Meghalaya, Mizoram and Nagaland; liliums in Nagaland; rose in Mizoram, Meghalaya and Nagaland; Gladiolus in Tripura; carnation in Meghalaya and Nagaland; gladiolus in Tripura, Meghalaya, Nagaland and Manipur and gerbera in Meghalaya and Assam (Table 6). In gladiolus and gerbera, suitable varieties have been identified for the region.

Table 6: Potential flower crops of NEH Region

S.No.	States	Crops
1.	Assam	Orchids (dendrobium, vanda), anthurium, marigold, tuberose, jasmine, gladiolus, gerbera, bird of paradise, chrysanthemum
2.	Arunachal Pradesh	Gerbera, rose, carnation, gladiolus, orchids (cymbidium, paphiopedilum, vanda, oncidium), lilium, anthurium, foliage, succulent plants, begonia

3.	Manipur	Orchids (dendrobium), anthurium, roses, gerbera, carnation, Leather leaf ferns, marigold
4.	Nagaland	Orchids (dendrobium), calla lily, heliconia, bird of paradise
5.	Tripura	Dendrobium orchids, anthurium, gerbera, lilium, tuberose, marigold, gladiolus, chrysanthemum
6.	Mizoram	Anthurium, bird of paradise, rose
7.	Meghalaya	Anthurium, rose, carnation, asiatic and oriental lilies, calla lily, bird of paradise, heliconia, gerbera, golden rod, leather leaf fern
8.	Sikkim	Anthurium, orchids (cymbidium, phalaenopsis, oncidium, cattleya), alstroemeria, calla lily, carnation, bird of paradise, gerbera, chrysanthemum

Gladiolus: Friendship, Her Majesty, Summer Times, BisBis, White Prosperity, Novalux, Jester Gold, Oscar, PusaArchana, PusaShagun and Dhanvantari.

Gerbera: Piton, Monarch, Pink Elegance, RCGH-117, Alesmeera.

Integrated Nutrient Management

Fertilizer schedule management based on analysis of leaf and soil nutrient status through appropriate diagnostic tools, use of soluble organic fertilizers through drip irrigation system, improvement of soil fertility through addition of green manures, composts, liquid manures, cowdung, cowurine, vermicompost, vermiwash and use of biofertilizer like VAM (*GlomusfasciculataumG,mosseae*) and other phosphate solubilizers like microphos and phosphobactrin @200g/ tree/year enhances phosphorus availability and improve fruit yield and quality. *Glomusfasciculatum* enhances root growth and nutrient uptake in citrus, yield increase up to 20% in papaya, 20-30% in banana and 50% saving of super phosphate in chilli *and Glomusmosseae* saves 50% super phosphate in chilli and yield increase up 35% in tomato.

Integrated soil and water management

Mulching, preparation of half- moon terraces, bench terraces, contour bunding, planting of double row of herbaceous fruits crops across the slope and fruit trees on the contour line, use of drip irrigation system for high value fruit crops and construction of water storage tank (Jalkund) for life saving irrigation have been observed as proven technologies for soil and water management.

Development of horticulture based cropping system

Suitable land use systems such as Agrihorti, Agri-horti-silvi pastoral, mixed horti, pure horti, horti-silvi-patoral and multi-tier horticulture system should be developed based on agroclimate zones, crop priority, topography and socio-economical factors.

Importance of Organic Farming

Improper farming practices such as monocropping, imbalanced fertilization, poor soil organic matter management, soil contamination, soil compaction, mining of soil nutrients, water logging, depletion of ground water, decline in soil biodiversity and changing pest and disease complex and application of imbalanced NPK fertilizers ratio of 7.9:3:1 as against normal values of 4:2:1 are the major factors for soil degradation. Looking the adverse effects of fertilizers and chemicals stress is being given to promote organic farming. The data indicates that per consumption of fertilizers and pesticides in India is 91.5 kg and 0.38 kg, respectively which are far below than other countries. The unprecedented rise in population will lead to the increased demand of food. The projected population for 2020 is 154 crores for which there will be a requirement of 385 million tonnes of food grain (Bhattacharya, 2004). It is estimated that plant nutrients (NPK) addition during 2020 will be removal of 37.46 million tonnes nutrients by crops for which nutrient additions generally fall short of requirement i.e. 7.86 million tonnes. In this context, the projected (2025) availability of plant nutrients trapable from organic sources is 7.75million tonnes. Besides, there is growing demand for organic produce which gets higher remunerative price even if yields are lower. Most of people believe that organic farming is the right choice for the long term future of the earth.

In India, 74% farmers owning less than 2 ha of land can adopt organic practices easily and manage farm inputs and labour more efficiently.The average productivity of 1.0 t/ha of Traditional rainfed agriculture can be enhanced upto 2.0 t/ha through organic agriculture.There is tremendous potential to substitute organic fertilizers through organics mainly crop residues, green loppings, green manures, organic manures, oil cakes, bio-fertilizers, crop rotations, animal excreta etc.The existence of traditional knowledge and farming systems have strong linkage between agriculture, livestock and others which can provide better livelihood security to farm families.Varied agro-climatic conditions and longer growing periods are prevailing in India to meet the offseason growing demands of export and the domestic markets.The organic sector is growing at 25-30%, creating employment opportunities for production, processing and value chain at rural areas and also marketing in cities.

Present Status of Organic Agriculture

Organic agriculture is gaining importance in European countries, USA, Canada, Australia, Japan, China, India and others. In the world, 160 countries are practicing organic agriculture with a certified organic cultivated area of 37 million ha and non-agriculture organic area (wild harvest) of 41.9 million ha. India has the largest number certified organic producers over 67,725 with 3.95 million ha under organic farming.During 2010, India accounted for 1,624,339 MT of certified organic produce. This country is the largest producer of organic cotton. Karnataka the first state to introduce policy on organic farming on 24-03-2004 and organic movement in India. In India, Uttarakhand, North-East States, Chhatishgarh have taken initiatives for promoting organic farming.

North East States are selected because the land is almost virgin and the crops are virtually organic. The use of inorganic fertilizers and chemicals is meagre in the region. All the households are maintaining livestocks producing sufficient quantities of on-farm manures. The region is receiving very high rainfall leads to production of biomass including weeds, shrubs, and herb which could be efficiently used in organic production. The region has the potential of about 47 mt of organic manure including 37mt from animal excreta and 9 million tons from crop residues. The region is home to some niche crops like Assam lemon, Joha rice, medicinal plants and passion fruits. North Eastern Region (NER) accounts for 45% of total pineapple production in India. Sikkim is the largest producer of large cardamom in the world. NER is the fourth largest producer of oranges in India. Extent of chemical consumption in farming is less than the national average. Besides, eighteen lakh ha of land in NER can be classified as 'Organic by Default'. In January, 2016 , Sikkim has been declared as "First Organic State of India".

Organic Market

Major organic produce dominate exports for developing countries include plantation commodities viz. tea, coffee, spices, high value fresh and processed fruits and vegetables, oilseeds, cotton, , cereals, pulses and meat/poultry and fish products. In 2000, Ministry of Commerce, Government of India, launched the National Programme for Organic Production (NPOP) to ensure focused and well directed development of agriculture. Recently, the following agencies have been approved as accreditation agencies by the Government of India to formulate the National Program for Organic Production.

- Agricultural and Processed Food Products Export Development Authority (APEDA)
- Spice Board
- Coffee Board
- Tea Board
- Coconut Development Board
- Directorate of Cashew and Cocoa Development

Under NPOP, documents such as national standards, accreditation criteria for accrediting inspection and certification agencies, accreditation procedure, inspection and certification procedures have been prepared on the basis of guidelines evolved by International Organizations viz., IFOAM, EU regulations and CODEX Standards.

Organic Sources and their Role

The potential organic sources of plant nutrients are green manure crops, crop rotation, crop residues, organic manures, FYM, Night soil, sludges, oilcakes, blood meal, compost, phospho-compost, vermin-compost, biogass slurry, agricultural wastes, press mud, Biodynamic preparations, biofertilizers etc. (Sharma, 2004). Organic sources improve soil structure, soil aeration, water holding capacity and reduce soil losses due to erosion. They supply nutrients in a balanced ratio and stimulate soil flora and fauna.

Organic sources for essential elements (De *et al.,* 2007)

Essential elements	Sources
Phosphorus (P)	Poultry litter, rock phosphate
Potassium (K)	Cover crop, mined granite, basalt, feldsper
Calcium (Ca), Magnesium(Mg) and Sulphur (S)	Sea weed extracts, dolomite, gypsum, limestone, crab shells
Micronutrients (B,Cu,Fe, Mn,Mo and Zn)	Liquid sea weed extracts, rock powder

1 ton green leaf manure= 10 kg of urea
Optimum dose of green biomass = 4 to 5 tonnes/ha

Traditional Additives for Organic Agriculture

Traditional additives comprises of bulky organic manures mainly Farm Yard Manure (FYM) for improvement of soil total nitrogen (Bharadwaj and Guar, 1985); compost which can be prepared from crop residues, weeds and vegetative biomass, sugarcane trash and pressmud, coir waste, tea waste; green manures obtained from sunhemp, *Sesbaniaaculeata,* cowpea, cluster bean; sewage and sludge from cities and towns; sheep, goat and poultry manure; concentrated organic manures viz. oilcakes like castor cake, karanj cake, cotton seed cake, mahua cake, safflower cake, groundnut cake, linseed cake, rapeseed cake and sesame cake; meat meal, blood meal and fish meal which have adequate amounts of NPK (0.5-2.0% N, 0.2-1.0% P_2O_5 and 0.5-2.0% K_2O) and higher C: N ratio.Neem cake is widely used for controlling nematode in oranges and tomato.

Adoption of Integrated Farming Systems

Integrated Farming Systems aim to increase production, income and to improve nutrition of small scale farmers with available resources. Components of Framing system includes all types of crops like field crops, horticultural trees, animals (cattle, buffalo, pig, goats, sheep, fish), sericulture, apiculture, mushroom cultivation, forage crops, flowers and vegetable cultivation, biogas production and non-conventional plant resources. An effective farming system include the cropping system (s), the livestock system (s) and the farm household. Diversified farming system increases productivity, profitability, balanced food ratio, adoption of new technology; solves energy, fodder, fuel and timber crisis; avoids degradation of forests and environmental pollution, generates employment and provides opportunities for Agro-industries by utilizing the natural resources efficiently and this will help in maintaining sustainability (Table 7).

Table 7: Main components of organic crop production

Diversification and integration of components	Sustainability	Natural Plant Nutrition	Natural Pest Management
Crop rotation	Crop rotation	Legumes in crop rotation	Diverse crop rotation
Animal manures	Efficient cropping system	Composting	Green manure
Composting	Cover cropping	Cowdung slurry	Cover crops
Intercropping	Intercropping	Green manure	Guard crops
Mulching	Biocontrol	Vermi-composting	Break cropping
Farm scapping	Mulching	Farm litter management	Use botanical extracts
	Resource conservation	Crop residues management	Weed and pest smothering crop
	Efficient recycling		Fire
			Sanitation
			Tillage
			Biological traps

Important Substrate Media for Ornamental Plants

Cocopeat: Cocopeat is a multi purpose growing medium made out of coconut husk. The fibrous coconut husk is pre washed, machine dried, sieved and made free from sand and other contaminations such as animal and plant residue. Cocopeat is a very good alternative to traditional peat moss and Rock wool. It has air filled porosity and high water holding capacity and is an ideal growing medium for the plant crops. It is 100% organic and eco- friendly, free from soil borne pathogen and weed. It has a pH of 5.7 – 6.5 , EC level <1 mS/cm is ideal for plant growth.

Vermiculite: This is a micacious mineral produced by heating to 745°C. These are expanded plate like particles having very high water holding capacity, aeration and drainage. Vermiculite is hydrated aluminium iron magnesium silicate material with accordion like structure. It has a very low bulk density, a neutral pH, a high CEC and small amounts of potassium and magnesium. It is well suited for propagation media.

Perlite: This is a siliceous mineral of volcanic origin. Perlite is rigid, sterile and essentially infertile with minimum CEC and neutral pH. Perlite may be included in the mix to increase aeration and to lower bulk density. It is generally recommended for use in a propagation media.

Rock wool: It is produced by burning a mixture of coke, basalt, lime stone and slag from iron production at 1600^{0}c temperature. The fiber rock wool is available in cubes and slabs. It is non-biodegradable, pH is 7 to 8.5, no buffering capacity, contain calcium, magnesium, iron, manganese, copper and zinc, CEC is negligible, does not require pasteurization, light is weight, high water holding capacity and good aeration.

Peat:It is the common component of artificial growing media. Peat are composed of several species of plant including mosses, sedges and grasses. In a growing medium, the value of peat is determined by the type of plant material and degree of decomposition. Peat are classified into four categories *viz*, Hypnaceous moss, Reed and Sedge, Humus or Muck and Sphagnum Moss.

Humus or Muck: These peat are the decomposed debris of finely divided plant materials of unknown origin. It contains sufficient quantities of silt and clay particles and does not improve drainage or aeration. Humus is not recommended for use in growing media due to its rapid decomposition and particle size.

Sphagnum Moss: This is the dehydrated remains of acid bog plants from the genus Sphagnum. It is low in soluble salts, long lasting in the mixture, uniform in the composition and improves drainage and aeration. Sphagnum peat moss has a good water holding capacity, high CEC, low nutrient levels and a comparatively low pH (3.0-4.5). It is the most desirable form of organic matter for the preparation of growing media.

Bark: These are by-products of the pulp, paper and plywood industries. Hardwood bark is the common ingredient of a growing medium. Bark is aged with lime and leaching to reduce the risk of toxicity of plants. Hardwood bark along with nitrogen makes a good potting media. Soft wood bark is acceptable but it lowers the pH of a media and so, liming is required. Barks are lighter in weight, sterile and have capacity to retain water as well as to drain the same.

Vermi-compost: It is prepared from the organic wastes upon the action of earthworms. It contains 2.5-3.0% Nitrogen, 1.0-1.5% Phosphorus and 1.5-2.0% Potash.

Vermiwash: It is washings from the earthworms collected during the preparation of vermin-compost, used as spray in raising of nursery, lawn and orchids (Ismail and Pramoth, 1995). Vermiwash is rich in growth promoting substances.

FYM/ Compost: It is prepared from the decomposition of organic wastes through anaerobic organisms. It contains fair amount of macro and micro-nutrients and most commonly used organic supplement given to the flower crop cultivation. FYM contains 0.5-1.5% N, 0.4-0.8% P_2O_5 and 0.5-0.9% K_2O whereas Garden compost contains 0.5% N, 0.3% P_2O_5 and 0.8% K_2O.

Panchgavya: It is a natural growth promoter and contains essential plant nutrients. It is prepared by mixing of fresh cowdung (5kg), cow's urine (3 litres), cow's milk (2 litres), cow's curd (1 litre), cow's ghee (100g), sugarcane juice (3 litres), Tender coconut water (3litres), banana fruits (12 Nos.). Usually, 1 to 5% solution of Panchgavya at 15 days intervals is used only after filtering.

Coconut water/Liquid manures: Tender coconut water contains growth promoting substances such as cytokinin which is reported to increase vaselife of cut flowers and is also used in the tissue culture media. Liquid leaf (ground fern + *Artemisia vulgaris*) and cowdung manures are rich in various anions (Table 8) and cations (Table 9) and used as foliar spray for cultivation of commercial flowers.

Table 8: Anions in liquid leaf and cowdung manures estimated by ion chromatography

Anion concentration (ppm)	Liquid manure: water(1:30)	Liquid manure: water(1:20)	Liquid manure: water(1:10)	Liquid manure: water (1:5)	Tape water (1:10)	Coconut water
Flouride	14.550	9.20	4.18	2.220	0.393	18.428
Chloride	91.71	58.40	29.61	17.835	3.230	2336.982
Bromide	15.848	10.78	5.43	3.136	0.285	4.884
Nitrate	101.70	66.43	33.48	15.308	0.312	—
Phosphate	174.924	122.79	64.74	53.746	0.238	351.182
Sulfate	939.069	615.21	308.03	138.693	14.825	86.558

Table 9: Cations in tape water and coconut water estimated by ion chromatography

Cation concentration (ppm)	Coconut water (1:10)	Tape water
Sodium (Na)	166.573	—
Potassium (K)	1816.479	30.321
Calcium (Ca)	254.847	—
Magnesium (Mg)	135.442	5.115

Traditional organic cultivation

1. Agnihotra farming: Growth promoter, insecticide
2. Panchagavya: Quality banana production, yield increase in cole crops.
3. Cowdung: Insect pests & disease control, seed germination.
4. Tobacco leaf extraction: Insect control.
5. Neem or karaglate: Soil borne disease control.
6. Alder agriculture in Nagaland: Growing of colocassia, pumpkin, chilli. Alder agriculture in Sikkim: Large cardamom.
7. Intercropping/mixed cropping: Rice + Cucurbits

   Rice+ Banana. A.P.

   Maize+ French bean
8. Temperate Horticulture in Sikkim: Apple+Potato Radish, Cabbage.

   Mid hill Horticulture in Sikkim; Mandarin+ Ginger+ Sweet potato.
9. Contour trench farming in Mizoram: Pineapple planting.

Principle and Practices of Organic Horticulture

Marketing and economic consideration

- Protection, volume, size, quality and reliable marketing strategy.
- Processing of unsuitable fruits for fresh markets to make up production cost.
- To meet buyers expectation of volume, quality and timing.
- Cost-benefit analysis of markets.

Planning and planting of an organic orchard

- Site selection based on topography, soil type, water filtration rate, frost pattern, maximum and minimum temperature, length of growing season, rainfall pattern, availability of irrigation water, possibility of water table and wind velocity.
- Crops and varieties suited for the region.
- Crops species selection.
- Small size short-term investment: Strawberries, Kiwi, grapes, vegetables, ornamentals.
- Big size long term investment; Apple, peach, orange, papaya, pears, sapota Selection of variety: based on

- Early, mid, late;
- Adaptability, chilling requirement
- Water requirement, dwarf, semi dwarf or standard
- Resistance to insects pests and diseases.

Procurement of planting materials of organic stock from reputed nurseries a heady certified by inspectors.

Site preparation through

- Adjustment of soil P^H through soil test
- Nutritionally balanced soil with addition of organic matter.
- Well planned sequence of tillage
- Soil solarization by using plastic film on moist soil for 1 to 2 months.

Orchard layout and design

Based on topography, species, stature, maturity time, weed prevalence.

Orchard management

- Covering cropping using grasses or legumes.
- Crop rotation with annuals or perennials.
- Pest management through IPM.
- Nutrient management through compost mixing with soil using shallow tillage, addition of organic manures like leaf manure, cowdung, vermicompost, biodynamic field sprays, cow pat pit manures, liquid manures, hybrid compost, vermiwash, biofertilizers etc.
- Organic weed management through cover crops, organic mulches (straw, spoiled hay, leaves, wood dust, sugarcane trash etc.), use of geotextiles, sheet mulch, use of weed harrowing, Implements or heel hoe, use of allowed herbicides like vinegar, citric acid in each terraces.
- Water management through drip irrigation, half moon terracing, full moon terracing; planting across the slope, vegetable barriers etc.

Post harvest management starting from washing, grazing, packing and storage should be practiced carefully to prevent contamination.

Biofertilizers: These are ready to use live formulates of beneficial microorganisms which on application to seed, root or soil fix atmospheric nitrogen or solubilize/ mobilize plant nutrients or otherwise stimulate plant growth substances. These free living or symbiotic microorganisms directly or indirectly contribute nutrition to crop plants. Free living organisms fix nitrogen without any association with any other organisms e.g. *Azotobacter*. Symbiotic association such as Legume-Rhizobium symbiosis and Anabaena-Azolla symbiosis expresses the mutual beneficial partnerships between the two organisms. Associative symbiotics fix nitrogen in C4 plants and the organisms responsible for this process is *Azospirillum*. Based on type of microorganisms, the biofertilizer can also be classified as Bacterial biofertilizers (*Rhizobium*, *Azospirillum*, *Azotobacter*, *Phosphobacteria*), Fungal biofertilizers (Mycorrhiza, Algal biofertilizers (Blue Green Algae, Azolla), and Actinomycetesbiofertilizers (Frankia). Legumes on an average contribute 50-200 kg N/ha whereas biofertlizers containing microorganisms like *Rhizobium*, *Azotobacter*, *Azospirillum*, Blue Green Algae, Azolla may contribute 40-60 kg N/ha and P-solubilizers may add 20-30 kg P_2O_5 /ha.

Azospirillum excretes growth promoting hormone (IAA) which results in more root biomass. It promotes uptake of nutrients like NO_3, NH_4, H_2PO_4, K and Fe. In flowers and ornamental plants, seedlings are treated with 1.5-2.0kg Azospirillum inoculants per hectare for getting beneficial effect. Azotobacter is well known to synthesize IAA, Gibberellins and B-vitamins and it has fungistatic properties. In flowers, seedlings treated with 1.5-2.0 kg Azotobacter inoculants per ha whereas bulbs/tubers are treated with 3.0-4.0 kg/ha for getting better effect (Table 10).

Table 10: Method and rate of application of biofertilizers in Horticultural crops

Crops	Type of biofertilizers	Method of application	Quantity required/ha
Peas, beans	Rhizobium	Seed treatment	1.0-2.0 kg
Cowpea	Rhizobium	Seed treatment	0.4-0.6kg
Radish, Spinah, Okra	Azospirillium	Seed treatment	0.4-0.6 kg
Onion, Brinjal, Tomato, Chilli, Cabbage	Azospirillium	Seedling treatment	1.5 to 2.0 kg
Flowers	Azospirillium	Seedling treatment	1.5 to 2.0 kg
Sugar beet	Azotobacter	Seed treatment	2.0-3.0 kg
Potato	Azotobacter	Soil/tuber treatment	4.0-5.0 kg
Radish, spinach, okra	Azotobacter	Seed treatment	0.4-0.6 kg
Turnip, carrot	Azotobacter	Seed treatment	0.2 kg
Onion, brinjal, cauliflower, cabbage, tomato, chilli	Azotobacter	Seedling treatment	1.5 to 2.0 kg
Flowers	Azotobacter	Seedling treatment	1.5 to 2.0 kg
	Azotobacter	Bulb/tuber treatment	3.0-4.0kg
Fruit trees	Azotobacter	Soil treatment	8-10g /tree
Vegetables (nursery)	VAM	Soil treatment	100g/m2
Polybag raised crops	VAM	Soil treatment	5-10g/packet
Fruit trees	VAM	Soil treatment	100-200g/tree

Organic Substrate Media for Growing various Flower Crops

Compost, garden soil, carbonized rice hull, coir dust, saw dust, fine sand, poultry manure and other organic materials are used as potting materials for various ornamentals (Table 11). Potting media should be sterilized in boiling water, solar irradiation or by burning or heating before planting to control soil borne diseases.

Table 11: Organic substrate media for growing various ornamentals (De, 2013; Rajiv Kumar and L.C. De, 2007)

Name of ornamentals	Substrate media
Rooted cuttings	Sand, loam, leaf mould
Hard wood plants	Sand, loam, peat moss, leaf mould, humus, cowmanure
Rose	Cowdung, loam, sand
Bulbous ornamentals	Cowdung, loam, sand, leafmould, charcoal
Terrestrial orchids	Leaf mould, sand, caly soil, bone meal, saw dust, wood charcoal dust, wood shavings, manures
Epiphytic orchids	Charcoal, brickpieces, sphagnum moss, tree bark
Anthurium	Coirpith, coarse sand, brick pieces/tiles, sugarcane baggase, leaf mould, rockwool, sawdust, tree barks, wood shavings, poultry manures
Gerbera	Sand, coir pith/cocopeat, leaf mould, FYm, vermicompost
Foliage & potted plants	FYM, sand, leaf mould, cocopeat, Charcoal
Carnation	Peat moss, perlite, sand
Lilium	Agropeat, vermicompost
Chrysanthemum	Cocopeat, garden compost
Cymbidium orchids	Cocochips, cocopeat, brick pieces, vermiculite, osmocot, dry leaf fern
Dendrobium orchids	Cocochips, brick pieces, tree barks
Vanda orchids	Cocochips, brick pieces, tree barks, dry leaf fern
Phalaenopsis orchids	Cocochips, brick pieces, leaf mould, green moss, sphagnum moss
Oncidium orchids	Cocochips, brick pieces, leaf mould
Cattleya orchids	Cocochips, brick pieces, leaf mould, dry leaf fern

Biopesticides and Biocontrol Agents for Insect-pests and Disease Management

There are several ways to control insect-pests in eco-friendly way with bio-control methods. Some of important tactics are

Cultural practices: crop rotation, use of resistant varieties, mixed cropping, habitat manipulation, trap crops.

Physical and mechanical methods: Removal of infected/infested plants/plant parts, removal of visible eggs, larvae etc., use of light traps, use of pheromone traps, use of colour stripes etc.

Biological control: Biological control is one of the important ways to manage insect pests and diseases in organic farming by using naturally occurring biological agents.

Predators: Biological agents which eat insect pests, e.g., frogs, lizards, ducks, spider, dragon fly, coccinellid beetle, meadow grass hoppers, mired bugs, crickets, ants, sparrows and other birds.

Parasitoids: These are friendly species which complete their life cycle on different stages of insect-pests. There are different types of parasitoids including egg parasitoids, larval parasitoids, pupalparasitoids, adults, egg-larval parasitoids, larval-pupalparasitoids etc., e.g., *Trichogramma* spp., *Bracon* spp., *Apanteles* spp, *Brachymeria* spp. An encyrtid wasp, *Anagyrus* sp (Howard) as a parasitoid of long tailed mealy bug, *Psuedococcus longispinus* and Mealybug, *Pseudococcus* sp. infesting Orchids from Sikkim, India.

Microbes: These microorganisms cause various diseases in the insect-pests. Fungi like *Beauvaria* spp., *Trichoderma* spp., *Metarrhizum* spp., *Entomophora* spp. act as biocontrol agents. Among viruses, Nuclear Polyhedrosis virus, Granulosisvirus arebiocontrol agents. Bacteria like *Pseudomonas* spp., *Bacillus thuringiensis*, *Bacilluspopoli* act as biocontrol agents. Microbial biopesticides, *Metarhiziuma nisopliae* and *Verticillium lecanii* @ 2 ml/Litre of water was found effective in reducing the crawlers populations of Biosduval scale insect, *Diaspis biosduvalii* with mean population reduction of 73.70% and 81.35%, respectively.

Naturally Plant Products: These are botanical pesticides derived from plants, e.g., pyrethrins, nicotenes, azadirachtin, rotenone etc. Highest mean population reduction in mite population was observed in plants treated with 5 % extracts of *Allium sativum* and neem oil (0.03%), respectively.

Organic Farming in North East Region

North Eastern States where fertilizer consumption is less than 25 kg/yr are suitable for organic farming. The region is rich in biodiversity and blessed with gene centers for citrus, banana, cucumber, brinjal, 3000 numbers of indigenous crop germplasms; 900 species of orchid, 9 genera of medicinal plants and five genera of aromatic plants.

North East States are selected because the land is almost virgin and the crops are virtually organic. The use of inorganic fertilizers and chemicals is meagre in the region. All the households are maintaining livestocks producing sufficient quantities of on-farm manures. The region is receiving very high rainfall leads to production of biomass including weeds, shrubs, and herb which could be efficiently used in organic production. The region has the potential of about 47 mt of organic manure including 37mt from animal excreta and 9 million tons from crop residues. The region is home to some niche crops like Assam lemon, Joha rice, medicinal plants and passion fruits. North Eastern Region (NER) accounts for 45% of total pineapple production in India. Sikkim is the largest producer of large cardamom in the world. NER is the fourth largest producer of oranges in India. Extent of chemical consumption in farming is less than the national average. Besides, eighteen lakh ha of land in NER can be classified as 'Organic by Default'. In January, 2016 , Sikkim has been declared as "First Organic State of India".

Priority Crops	Pineapple (Assam, Meghalaya, Tripura)
	Passion fruit (Mizoram, Nagaland, Manipur, Sikkim)
	Kiwi (Sikkim, Arunachal)
	Apple (Arunachal, Sikkim)
	Orange (Assam, Meghalaya, Sikkim, Arunachal)
	Radish (Meghalaya)
	Ginger (Meghalaya, Mizoram, Sikkim, Nagaland)
	Turmeric (Meghalaya, Assam, Mizoram)
	Large cardamom (Arunachal, Sikkim)
	Potato(A.P.,Meghalaya)
	Colocasia(Assam, Meghalaya)
	Orchid(Sikkim, Arunachal)
	Gladiolus(Meghalaya, Tripura, Nagaland)
	Blackpepper, Arecanut(Assam, Tripura)

Experimental results under organic cultivation

Crop	Nutrient requirement	Yield
Cabbage	Poultry manure10t/ha+Panchagavya	460 t/ha
Cauliflower	FYM (20t/ha)	348 t/ha
	Poultry manure (10t/ha)	342 t/ha
	FYM + Panchagavya	333 t/ha
Broccoli	Poultry manure@10t/ha	190 t/ha
	FYM (20t/ha)	182 t/ha
Banana	Vermicompost +75% RDF +AM fungi	32.41 t/ha
Pomegranate	Black polythene mulch +full moon terracing	13kg/plant
Guava	Phosphobactrin	154g fruit weight

References

Bhardwaj, K.K. and Guar, A.C. (1985). Recycling of organic wastes, ICAR, New Delhi, pp.104.

Bhattacharya, P. 2004. Organic food Production in India.Agrobios India, Behind Nasrani Cinema, Chopasani Road, Jodhpur.

Chakraborty, M., M. Dutta, L.C. De, S. Mitra, K.R. Dhiman and N.P. Singh, (2006). Effect of organic sources on vegetable productivity with special reference to soil organic carbon, mineralization, nutrient availability and uptake in Tripura. In: Horticulture for Sustainable Income and Environmental Protection (*Advances in Vegetables, Spices, Plantation crops and Medicinal and Aromatic plants*). pp.501-509. Concept Publishing Co. New Delhi-59.

De, L.C. (2013). 'Nursery and Landscaping', pp. 248 Published by Pointer Publisher, Jaipur, Rajasthan.

De, L.C., Wilson Rai, SumanThapa and Singh, D.R. (2016). Input Management in Organic Floriculture- an Overview. *International Journal of Humanities, Arts, Medicine and Sciences* (BEST: IJHAMS), Vol. 4, Issue 5, May 2016, 47- 56.

De, L. C., Sanwal, S.K., Akath Singh, Babu, K.D., Rajiv Kumar and Yadav, R.K. (2007). Organic Input Management in Important Horticultural crops of NEH Region. In:'Advances in Organic Farming Technology in India'(eds.) G.C. Munda, P.K. Ghosh, Anup Das, S.V. Ngachan and K.M. Bujarbaruah); Published by the Director, ICAR Research Complex for NEH Region, Umiam-793103, Meghalaya, pp.585.

De, L.C. and Sarangi, S.K. (2006). ITK Used in Organic Agriculture, pp 423-438 In: '*Trends in Organic Farming in India*', S.S. Purohit and D. Gehlot (Eds); 438p. Agrobios (India), Agro House, Jodhpur-342 002.

Kumar, Rajiv and De, L.C. (2007). Organic Floriculture-Potential and Prospect. In: 'Advances in Organic Farming Technology in India' (eds.) G.C. Munda, P.K. Ghosh, Anup Das, S.V. Ngachan and K.M. Bujarbaruah); Published by the Director, ICAR Research Complex for NEH Region, Umiam-793103, Meghalaya, pp.585.

Ismail, A.A. and Pramoth, A. (1995).Vermiwash, a potent bio-organic liquid 'ferticide'. In: Ismail Vermicology. The biology of earthworms. Orient Longman, pp. 92.

Sanwal, S.K., Yadav, R.K. and De, L.C. (2007). Natural Growth Promoter-an alternative to Chemical fertilizers and pesticides in organic horticulture. In: 'Advances in Organic Farming Technology in India'(eds.) G.C. Munda, P.K. Ghosh, Anup Das, S.V. Ngachanand K.M. Bujarbaruah); Published by the Director, ICAR Research Complex for NEH Region, Umiam-793103, Meghalaya, pp.585.

Sarangi, S.K. and De, L.C. (2006). Cultivation practices of organic turmeric and ginger in NEH Region, pp.380-395 In: '*Trends in organic farming in India*', S.S. Purohit and D. Gehlot (Eds); 438p. Agrobios (India), Agro House, Chopsani Road, Jodhpur-342 002.

Sharma, A.K. (2004). A Handbook of Organic Farming. Agrobios India, Behind Nasrani Cinema, Chopasani Road, Jodhpur.

Singh, H.P. (2001). Organic farming approach for sustainable coconut cultivation. *Indian Coconut Journal*, 3-7.

4

Humic Acid and Its Role in Vegetable Production

***T. Thomson*[1] *and T. Tejaswini*[2]**

[1]*College of Horticulture, Dr. YSR Horticultural University Venkataramannagudem - 534101, Tadepalligudem, West Godavari Andhra Pradesh*

[2]*College of Horticulture, Sardarkrushinagar Dantiwada Agricultural University Jagudan, Gujarat*

Abstract

Importance of organic sources of nutrients was recognized in current scenario in order to get higher yield without disturbing soil health. In this context integrated use of inorganic and organic sources of nutrients in crop production is becoming very crucial for assurance of food security on sustainable basis, which in turn not only improve the soil fertility for sustained crop productivity but also to reduce the cost of inorganic fertilizers. Among the various organic sources humic acid plays a major role in the sustainable production of vegetables. Humic acid is a natural bio-stimulant that is derived from leonardite and is among the most concentrated organic material available today. Humic acid is an organic chemical produced by decaying plants and animals residues by a process called Humification. Humic acid is not only increases the yield but also improves the soil fertility, soil productivity, soil structure and texture, promoting plant growth and resistance against various pests, diseases. In this context use of humic acid as organic sources of nutrients in crop production is becoming very crucial for assurance of food security, sustainability and improving soil health.

Keywords: *Humic acid, Humus, Vegetables, Yield, Quality*

Introduction

Organic cultivation is the form of agriculture that relies on techniques such as crop rotation, green manure, compost, and biofertilizers which helps to maintain soil productivity. Organic farming excludes or strictly limits the use of chemical fertilizers, pesticides (herbicide, insecticides and fungicides), plant growth regulators, live stock feed additives, antibiotics and genetically modified organisms. Organic farming is a holistic production management system which promotes & enhances health of agro ecosystem related to biodiversity, nutrient bio-cycle, soil biology & microbial activities. It is a system of farming without the use of chemical inputs like fertilizers or insecticides etc and is primarily based on principles of use of natural on farm organic inputs (like FYM ;Green manures; oil cakes, Humic acid, cow urine cow dung, vermicompost etc.) & also natural biological pest control & plant protection measures to promote agro economic system & soil biological activity. India is primarily agrarian country and this sector provides livelihood to a major part of the population. The indiscriminate use of agro chemicals since green revolution resulted in adverse effect on the soil fertility; crop productivity; quality of produce and more specifically on the environment system.

Thus, the importance of organic sources of nutrients was recognized in current scenario in order to get higher yield without disturbing soil health. In this context integrated use of inorganic and organic sources of nutrients in crop production is becoming very crucial for assurance of food security on sustainable basis, which in turn not only improve the soil fertility for sustained crop productivity but also to reduce the cost of inorganic fertilizers.

Vegetables are defined as those annual, biennial or perennial herbaceous plants of which succulent parts are commonly used for culinary purposes and of which the stem, root, flowers, fruits, petiole or leaves are eaten cooked or raw as a main part of a meal, side dish, or appetizer. Vegetables are called protective food as their consumption can prevent several diseases and are cheaper source of minerals, vitamins and fiber with high calorific values. As per ICMR, per capita consumption of vegetable is 350g/person/day (120g GLV's, 100g roots, 130g other vegetables). But the availability of vegetables in India is 210g/person/day. Vegetable growing is the most remunerative enterprise as it is adopted on small and marginal holding with high production in short duration. Being a source of farm income, it creates impact on the agricultural development and economy of the country. Vegetables are cheaper source of minerals, vitamins and fiber with high calorific values. There is an increasing demand of vegetables both for domestic as well as for export, which can earn valuable foreign exchange for country. In India vegetables occupies an area of about 939600ha with the annual production of 162897000MT & productivity of 17.3 MT/ha (NHB 2014-2015).

Humic acid is a natural bio-stimulant that is derived from leonardite and is among the most concentrated organic material available today. Humic acid is an organic chemical produced by decaying plants and animals residues by a process called Humification. Humic acid is one of the major components of humic substances which is dark brown in colour and the major constituents of soil organic matter humus that contributes to soil chemical and physical properties and are also precursors of some fossil fuels. Humic acid can also be found in peat, coal, many upland streams and ocean water. One source of humic acids are the sedimentation layers referred to as Leonardite. These layers were originally deep in the earth's crust, but over many years have been exhumed to near-surface location. Humic acids are found in high concentration in these layers. Leonardite is organic matter, which has not reached the state of coal and differs from soft brown coal by its high oxidation degree, a result of the process of coal formation, and has no value as fuel. The decomposition of concentrated organic acids is a lengthy process taking millions of years in the natural environment. Plants are absorbing carbon dioxide from the atmosphere and using the sun's energy to build plant biomass. These plants feed insects and vertebrates. As plants and animals die they contribute their carbon back to the bottom of the bog. Over millions of years this cycle of organic matter is concentrated and compressed into layers in the earth.

Elemental analysis of humic acid has shown that that consists largely of carbon and oxygen (about 50% and 40% respectively) which also contains hydrogen (about 5%), nitrogen (about 3%), phosphorous and sulfur (both less than 1%). Humic acid is a complex of closely related macromolecules. These molecules range in size from less than 1000 to more than 100,000 daltons, with the lower mass representing the new material. Humic substances of the soil provide a good source of energy to living organisms of the soil, as they do not have access to energy produced through photosynthesis like terrestrial plants. Soil organisms such as algae, yeast fungi, bacteria, nematodes, mycorrhizae, etc. perform important functions such as improving the soil fertility and structure and promoting plant growth and resistance against various diseases.

Plant residues

↓

Break down or transformation by microorganisms

↓

Sugars, polyphones, amino acids, quinines.

↓

Fulvic acid + Humic acid + Ulmic acid + Humin

↓

Results Increasing intensity of colour
Increasing molecular weight
Increasing cation exchange capacity

Humic acid

Its is not soluble in water under very acidic conditions (p^H<2) but soluble at higher p^H levels. Humic acid is the major portion of extract from humus used in horticulture. Humic acid appears dark brown to black colour.

Fulvic acid

Is the portion that is soluble in water under all p^H levels. It is light yellow to yellowish brown colour.

Ulmic acid

Also known as Hymatomelanic acid. Soluble in alkaline p^H

Humin

Its is the remaining portion of humus which is in minor portion which is not soluble in water at any p^H.

Importance of Humanic Humic Substances

Humus is the only path to continued agricultural sustainability; otherwise the Carbon (Food for soil organisms) diminishes to zero and renders the addition of man-made synthetic nutrients pointless. Humic products are extremely effective in combating salinity issues that arise from heavy use of synthetic fertilizers and well water degradation. Feeding soil bacteria massive amounts of Carbon (their primary food source) along with the Energy and Oxygen necessary to respirate the food, stimulates massive root growth. Larger root system means more water and nutrients available for food production which Increases Yield.

Adding significant amounts of clean activated Carbon back into our soils is readily available through Humic products. They add tremendous amounts of "natural" food for soil organisms in a cost-effective and sustainable manner, which also increases the Cation Exchange Capacity of the soil. Water and Nutrient holding capacity is greatly enhanced with the addition of Humic products. Soil is better "conditioned" with the addition of Carbon through Humic products.

Humic acid's role in improving soil quality

- Detoxify the soils affected by heavy soils.
- It has the ability to chelate positively charged multivalent ions.
- Acts as a Soil conditioner. Improve soil structure. Increase caution exchange capacity of soil.
- Stimulates biological activity in soil.
- Enhance nutrients uptake and increases the content of humus in soil.
- Reduce the poisoning of pesticide residue as well as protects against soil from contamination of heavy metal ions as well as other hazardous issues.

Humic acid's Role in Improving Plants growth

- Humic acid stimulates microbial activity by providing the indigenous microbes with a carbon source for food, thus encouraging plant's growth and activity.
- Plant growth stimulant, promote root development and stimulates seed germination.
- Enhance nutrient uptake by combining nutrients and humic acids and keep a well-balanced nutrition.
- Fertilizer effectiveness promoter. Greatly boost the usage rate of nitrogen fertilizer. Keeps 60% of nitrogen easily offered and also 40% as slow-release.
- Boost the durability of crops. such as chilling, dry spell, pest, disease as well as toppling resistance.
- Promotes healthier, stronger plants and enhance appearance.
- Improves Potassium content which is an important element of chlorophyll, accelerate the photosynthesis.

The Economic Benefits Are?

- Humic acids chelate nutrient compounds, especially iron, in the soil to a form suitable for plant utilization. Thus, the nutrient supply of plants is optimized. Increases up to 70% in yield, accompanied by a reduction up to 30% in the use of fertilizers and pesticides, as well as better and healthier growth of green grass, ornamentals, agricultural crops and woods can be attained with the regular application of first-quality humic acids. Furthermore, water holding capacity of soils is increased considerable, which means that the use of water can be reduced substantially.

- Best economic results can be obtained in light and sandy soils poor in humus, as well as on re-cultivation fields. The diverse positive impacts of humic acids are to be observed particularly in such soils. This is true for almost all soils in dry and warm regions. As a result of the high mineralization rate of organic substances, providing these soils with stable humic acids is indispensable for the maintenance and improvement of soil fertility.
- Current scientific studies show that the fertility of soil is determined to a very large extent by the content of humic acids. Their high cation-exchange capacity (CEC), the oxygen content as well as the above average water holding capacity are the reasons for the high value of using humic acids for improving soil fertility and plant growth. The most important feature of humic acids lies in their ability to bind insoluble metal ions, oxides and hydroxides, and to release them slowly and continually to plants when required. Due to these properties, humic acids are known to produce three types of effects: physical, chemical and biological.

The Benefits of Humic Acid

Chemical Benefits

- Helps convert mineral elements to plant food
- Rich in organic / mineral substances
- Retains soluble fertilizers in root zone
- Stabilizes nitrogen applications
- Breaks down salt complexes

Physical Benefits

- Make soils more friable or crumbly–breaks up hardpan (dense layer of soil)
- Increases water holding capacity (up to 7 times)
- Breaks down crop residues
- Increases aeration of soil

- Builds organic matter.
- Reduces soil erosion.
- Stabilizes soil temperature.

Biological Benefits

- Increases germination of seed.
- Improves plant respiration.
- Stimulates root growth.
- Stimulates plant enzyme.
- Increases vitamin content of plants.
- Improves nutrient uptake.
- Stimulates growth at all stages.
- Acts as an organic catalyst.
- Feeds microorganisms that recycle nutrients and produce antibiotics.

The Ecological Benefits Are?

- The ecological benefits of humic acids are diverse and present profitable and effective solutions for environmental problems and the preservation of the environment.
- First of all, soils with a high content of humic acids are a guarantee for low nitrate leaching and for optimum nutrient efficiency. A well-developed root system, which is achieved by a high content of humic acids, prevents nitrate and pesticides mixing in with ground water. Furthermore, a low content of nitrate is an indicator and a prerequisite for appropriate "organic animal feeds". It happens very often that growers use more fertilizers than plants can take up. This leads to nitrate concentration in soil, which is later to be found in ground water.
- Secondly, humic acids reduce the over-salination problem in the application of water-soluble mineral fertilizers. Humic acids are able to decrease high salt contents in soils and thus the resulting toxicities. The NH_4-toxicity of fertilizers containing ammonia is greatly reduced, which is of particular importance to young plants. Generally, humic acids reduce root burning which comes about through excessive salt concentrations in soils after fertilization; in case of permanent high levels of salt in soils, these are reduced. Furthermore, when humic acids are mixed with liquid fertilizers, the undesirable odor is reduced.

- Thirdly, humic acids are an effective means to fight against soil erosion. This is achieved both by increasing the ability of soil colloids to combine and by enhancing root system and plant development.
- Humic substances in soils and sediments can be divided into three main fractions: humic acids, fulvic acids, and humin.
- The humic and fulvic acids are extracted as a colloidal sol from soil and other solid phase sources into a strongly basic aqueous solution of sodium hydroxide or potassium hydroxide
- Humic acids are precipitated from this solution by adjusting the pH to 1 with hydrochloric acid, leaving the fulvic acids in solution. This is the operational distinction between humic and fulvic acids.
- Humin is insoluble in dilute alkali.
- The alcohol-soluble portion of the humic fraction is, in general, named *ulmic acid*. So-called "gray humic acids" (GHA) are soluble in low-ionic-strength alkaline media; "brown humic acids" (BHA) are soluble in alkaline conditions independent of ionic strength; and fulvic acids (FA) are soluble independent of pH and ionic strength.
- Liquid chromatography and liquid-liquid extraction can be used to separate the components that make up a humic substance. Substances identified include mono-, di-, and tri-hydroxy acids, fatty acids, dicarboxylic acids, linear alcohols, phenolic acids, and terpenoids.
- Application of HA improves soil aggregation, structure, fertility, and moisture holding capacity, and increases microbial activity (Chen and Aviad, 1990;Sharif *et al.*, 2002), microbial population, and Cation exchange capacity (Marinari *et al.*, 2000).
- Studies have shown beneficial effects of HA such as increasing cell membrane permeability, oxygen uptake, respiration and photosynthesis, phosphate uptake, and root elongation. In particular, the photosynthetic efficiency and chlorophyll contents of *Lolium perenne* (rye-grass) were significantly increased by HA application (Russo and Berlyn, 1990).
- Humic acid also has direct cytokinin (Zhang and Ervin,2004) and auxin or gibberellin-like stimulatory effects (Pizzeghello *et al.*, 2001), along with indirect effect on plant metabolism (Piccolo *et al.*, 1991).
- Humic acid, extracted from cattle, food and paper-waste vermicomposts, increased root growth of peppers (*Capsicum annuum grossum*L. 'King Arthur') (Arancon *et al.*, 2003).

- Its substrate drenching or foliar application were equally effective in maintaining higher root fresh and dry weights in cucumber (*Cucumis sativus* L.),
- Humic acid improved plant resistance to environmental stresses (Ferrara *et al.*, 2006).
- Incidence of plant diseases can also be limited with the application of HA (Nakamura, 1996). Humic acid application was also beneficial for nutrient uptake, particularly uptake of N, P, K, Mg, Ca, Zn, Fe, and Cu by plants (Fagbenro and Agboola, 1993; Nikbakht *et al.*, 2008).

Features & Advantages

- Soil conditioner. Improve soil structure. Increase cation exchange capacity of soil.
- Stimulates biological activity in soil.
- Enhance resistance to anxiety for soil, especially decrease high salts in alkaline dirt's.
- Enhance nutrients uptake and increases the content of humus in soil.
- Reduce the poisoning of pesticide residue as well as protects against soil from contamination of heavy metal ions as well as other hazardous issues.
- Improves the physical and chemical qualities of soils.
- Plant growth stimulant, promote root development and stimulates seed germination.
- Enhance nutrient uptake by combining nutrients and humic acids and keep a well-balanced nutrition.
- Fertilizer effectiveness promoter. Greatly boost the usage rate of nitrogen fertilizer. Keeps 60% of nitrogen easily offered and also 40% as slow-release.
- Boost the durability of crops. such as chilly, dry spell, pest, disease as well as toppling resistance.
- Promotes healthier, stronger plants and enhance appearance.
- Potassium is an important element of chlorophyll, accelerate the photosynthesis.

Differences between Humic acid and Fulvic acid

S.no	Humic acid	Fulvic acid
1.	Molecular weight higher than Fulvic acid.	Lower molecular weight
2.	Brown to black colour	Golden brown or light yellow to yellowish brown in colour
3.	It is mostly preferred for soil application.	Used for both foliar and soil application.
4.	Humic acid transports the minerals to cell membranes and cell structures.	Fulvic acid can able to penetrate the leaf and even the mitochondria of a cll, hence makes it great for foliar sprays and root drenches.

Humic acid

Fulvic acid

Effect of Humic acid on the Different Crops

Salman *et al.*, (2005) studied the effect of interaction between watermelon varieties & Humic acid on the fruit quality and they concluded that among the varieties S. belle recorded significantly superior results for all the parameters. Among the concentrations 6L/F recorded significantly superior results for all the parameters. Among the interaction (6L/F x S. belle) recorded maximum average fruit diameter (26.97 cm), average fruit weight (5.51 kg), cortex thickness (1.68 cm), firmness (16.33), TSS (10.49) whereas interaction between (6L/F x Gizal) recorded maximum average fruit length (28 cm) while interaction between (6L/F x Aswan) recorded maximum dry matter (4.85%).

Salman *et al.*, (2005) studied the effect of interaction between watermelon varieties & Humic acid on the yield parameters and they concluded that among the varieties S.belle recorded significantly maximum total yield (25.3 t/fd) whereas maximum early yield (13.2 t/fd), maximum marketable yield (17.5 t/fd) and significantly minimum unmarketable yield (4.1 t/fd) was recorded with the variety aswan. Among the concentrations 6L/F recorded significantly superior results for all the parameters. Among the interaction (6L/F x S. belle) recorded maximum early yield (15.68 t/fd) and minimum unmarketable yield (4.23 t/fd) while interaction between (6L/F x Gizal) recorded maximum total yield (33.8 t/fd) and marketable yield (23.49 t/fd).

Sajid *et al.,* (2012) studied the survival percentage (%), neck height (cm) & plant height (cm) of onion cultivars as affected by humic acid and they concluded that among the varieties parachinar local recorded significantly maximum neck height (7.5 cm) and plant height (80.9 cm) whereas survival percentage was found non significant. Among the concentrations humic acid 2 kg/ha recorded significantly maximum neck height (7.6 m) and plant height (75.3 cm) whereas survival percentage was found non significant.

Sajid *et al.,* (2012) studied the bulb weight (g), yield plot (kg) & total yield ha^{-1} (tones) of onion cultivars are affected by humic acid and they concluded that among the varieties parachinar local recorded significantly maximum bulb weight (94.2 g), yield per plot (22.9 kg) and total yield (36.1 t). Among the concentrations humic acid 2 kg/ha recorded significantly maximum bulb weight (96.4 g), yield per plot (22.4 kg) and total yield (35.3t).

Sajid *et al.,* (2012) studied the effect of lignite humic acid and inorganic fertilizers on growth attributes of onion and they concluded that significantly maximum plant height (49.5 cm), number of leaves/plant (47.2), root length (11.2 cm) and maximum dry matter production (2110 kg) was observed with the application of treatment T_5 (100%NPK+20KgHA/ha SA) and these were statistically at par with the treatment applications of $T_{3,}$ $T_{4,}$ $T_{6,}$ T_7 and T_8 while minimum was recorded with the treatment application of T_1 control.

Sangeetha and Singa Ram (2007) studied the effect of lignite humic acid and inorganic fertilizers on yield of onion and they concluded that significantly maximum number of bulbs/plant (9.8), bulb girth (9.2 cm) and bulb yield (18.7 t) was observed with the application of treatment T_5 100%NPK+20KgHA/ha SA and these were statistically at par with the treatment applications of T4 (100% NPK+10Kg HA/haSA) and T7 (100% NPK+10Kg HA/ha SA+0.1% HAFS) while minimum was recorded with the treatment application of T_1 control.

Kumar *et al.,* (2015) studied the effect of various concentrations of humic acid on growth parameters in okra (*Abelmoschus esculentus*) and they concluded that significantly minimum days taken for full germination (15.33) was recorded with the treatment application of T_{15} (150ml HA) and it is statistically at par with the treatment applications of T_4 (40ml HA) and T_9 (90ml HA) while maximum plant height (208.94 cm), length of leaf (24.67 cm), width of leaf (22.87 cm), leaf area index (447.99 sq. cm) and number of leaves (31.80) was also observed with the treatment application of T_{15} (150ml HA) whereas number of branches was found to be non significant.

Kumar *et al.,* (2015) studied the effect of various concentrations of humic acid on yield parameters in okra(*Abelmoschus esculentus*) and they concluded that significantly minimum days taken to first flowering (31.67) was recorded with the treatment application of T_{15} (150ml HA) and it is statistically at par with the treatment applications of T_4 (40ml HA) and T_7 (70ml HA) while maximum number of flowers per plant (31.13), minimum days taken to first fruiting (37.33), number of fruits per plant (23.13 cm), fruit length (13.20 cm) and was also observed with the treatment application of T_{15} (150ml HA) while maximum fruit diameter (26.67) was recorded with the treatment applications of T_{15} (150 ml HA) and T_{16} (Control) whereas dry weight of leaves and dry weight of plants was found to be non significant.

Kumar *et al.,* (2015) studied the effect of various concentrations of humic acid on yield parameters in okra (*Abelmoschus esculentus*) and they concluded that significantly maximum total yield (122 q)was recorded with the treatment application T_{15}(150ml HA) of whereas it is statistically atpar with the treatment application of T_4 (40ml HA)and T_7(70ml HA) while maximum vitamin c (22.33 mg) was also recorded with the treatment application of T_{15} (150ml HA) and it is statistically at par with the treatment applications of T_8 (80ml HA) and T_9 (90ml HA).

Aminifard *et al.,* (2012) studied the effect of humic acid on fruit quality characteristics of hot pepper (*Capsicum annum* var. Red chilli) and they concluded that significantly maximum titrable acidity (34.6 g/L), total soluble solids (11.25 ^{0}brix) and vitamin c (117 mg) was recorded with the treatment application of HA4 (250 mg/kg) while minimum was recorded with the treatment application of control HA0 control.

Aminifard *et al.,* (2012) studied the effect of humic acid on capsaicin and carotenoids of hot pepper (*Capsicum annum* var. Red chilli) and they concluded that significantly maximum capsaicin (281.7 mg/kg), lycopene (225 mg/kg) and β-carotene (500 mg/kg) was recorded with the treatment application of HA2 (100 mg/kg) while minimum was recorded with the treatment application of control HA0 control.

Aminifard *et al.,* (2012) studied the effect of humic acid on fruit antioxidant compounds of hot pepper (*Capsicum annum* var. Red chilli) and they concluded that significantly maximum fruit antioxidant activity (85.7 %), carbohydrates (133.9 mg) was recorded with the treatment application of HA4 (250 mg/kg) whereas total phenolics (41.5 mg) was recorded with the treatment application of HA3 (175 mg/kg) whereas maximum flavonoide (192.3 mg) was recorded with the treatment application of HA1.

Ugur *et al.,* (2013) studied the effect of humic acid application on leaf length (cm) and leaf width (cm) in green vegetables and they concluded that significantly maximum leaf length in sorrel (16.40 cm) and cress (7.01 cm) was recorded with the treatment application of 0.8% HA whereas maximum leaf length in rocket (14.26 cm) was recorded with the treatment application of 0.4% HA while significantly maximum leaf width in sorrel (8.73 cm) was recorded with the treatment application of 0.2% HA, whereas maximum leaf width in rocket (5.49 cm) and cress (3.47.73 cm) was recorded with the treatment applications of 0.4% HA and 0.4% HA respectively.

Ugur *et al.,* (2013) studied the effect of humic acid application on yield values in green vegetables and they concluded that significantly maximum yield in sorrel (2317.57 g/m^2), rocket (2127 g/m^2) and cress (1436.73 g/m^2) was recorded with the treatment application of 0.8% HA while minimum was recorded with the treatment application of control.

Conclusion

Humic acid is the important organic sources of nutrients for improving the growth, yield and quality of many vegetables. These organic sources not only increases the yield but also improves the soil fertility, soil productivity, soil structure and texture, promoting plant growth and resistance against various pests, diseases. In this context use of organic sources of nutrients in crop production is becoming very crucial for assurance of food security, sustainability and improving soil health.

References

Aminifard, M. H., Aroiee, H., Azizi, M., Nemati, H. and Hawa Z.E. Jaafar. (2012). Effect of Humicacid on Antioxidant activities and fruit quality of hot pepper (*Capsicum annum.* L.). *Journal of Herbs,Species & Medicinal Plants*, 18(1):360-369.

Ugur, A., Demirtas, B., Caglar, S., Zambi, O. and Turkmen, M. (2013). Effect of Humic acid application on yield and quality in green vegetables. *24th International Scientific Expert Conference of Agriculture and Food Industry Proceedings.*

Sajid, M., Abdur Rab., Shah, S.T., Ibadullah Jan., Ibsanul Haq., Haleem, B., Zamin, M., Alam, R. and Zada, H. (2012). Humic acid affects the bulb production of onion cultivars. *African Journal of Microbiology Research.* 6(28):5769-5776.

Kumar, P., Rana, D.K., Singh, V. and Nasseruddin Shah, K.H. (2015). Effect of Humic acid on growth, yield and quality of okra (*Abelmoschus esculentus* L.) Cv. Arka Anamika under subtropical conditions of Garhwal Himalaya. *International Journal for Innovative Research in Science and Technology.* 1(8):151-153.

Salman, S.R., Abou-Hussein, S.D., Abdel-Mawgoud. and El-Nemr, M.A. (2005). Fruit yield and Quality of watermelon as affected by hybrids and humicacid application. *Journal of Applied Sciences Research.* 1(1):51-58.

Sangeetha, M. and Singaram, P. (2007). Effect of lignite humic acid & inorganic fertilizers on growth and yield of onion. *The Asian Journal of Soil Sciences*.2(1):108-110.

5

Vermiwash (The Best Organic Sources for Sustainable Vegetable Production)

T. Tejaswini[1] and T. Thomson[2]

[1]College of Horticulture, Sardarkrushinagar Dantiwada Agricultural University Jagudan, Gujarat
[2]College of Horticulture, Dr. YSR Horticultural University Venkataramannagudem - 534101, Tadepalligudem, West Godavari Andhra Pradesh

Abstract

India is primarily agrarian country and this sector provides livelihood to a major part of the population. The indiscriminate use of agro chemicals since green revolution resulted in adverse effect on the soil fertility; crop productivity; quality of produce and more specifically on the environment system. Organic cultivation is the form of agriculture that relies on techniques such as crop rotation, green manure, compost, and biofertilizers which helps to maintain soil productivity. Organic farming excludes or strictly limits the use of chemical fertilizers, pesticides (herbicide, insecticides and fungicides), plant growth regulators, livestock feed additives, antibiotics and genetically modified organisms. Thus, the importance of organic sources of nutrients was recognized in current scenario in order to get higher yield without disturbing soil health. In this context integrated use of inorganic and organic sources of nutrients in crop production is becoming very crucial for assurance of food security on sustainable basis, which in turn not only improve the soil fertility for sustained crop productivity but also to reduce the cost of inorganic fertilizers. Among the various organic sources, Vermiwash was found to be one of the best sources for sustainable vegetable production. Vermiwash is the organic fertilizer decoction obtained from units of vermiculture/vermicompost as drainage. It is the liquid extract collected after the passage of water through the different layers of worm culture unit. This decoction contains excretory products of earthworm secretions of the worm-the coelomic fluid oozing through the dorsal pores, mucus, enzymes secreted by the worm & micro organism, plant nutrients, vitamins & plant growth promoting substances which in-creases the resistance power of crops against various diseases and enhances the growth and productivity of crops.

Keywords: *Vegetables, Vermiwash, organic, growth, yield and quality*

Introduction

Vegetables play an important role in human nutrition. Vegetables are defined as those annual, biennial or perennial herbaceous plants of which succulent parts are commonly used for culinary purposes and of which the stem, root, flowers, fruits, petiole or leaves are eaten cooked or raw as a main part of a meal, side dish, or appetizer. Vegetables are called protective food as their consumption can prevent several diseases and are cheaper source of minerals, vitamins and fiber with high calorific values. As per ICMR, per capita consumption of vegetable is 350g/ person/day (120g GLV's, 100g roots, 130g other vegetables). But the availability of vegetables in India is 210g/person/day. Vegetable growing is the most remunerative enterprise as it is adopted on small and marginal holding with high production in short duration. Being a source of farm income, it creates impact on the agricultural development and economy of the country. Vegetables are cheaper source of minerals, vitamins and fiber with high calorific values. There is an increasing demand of vegetables both for domestic as well as for export, which can earn valuable foreign exchange for country.

In India vegetables occupies an area of about 939600ha with the annual production of 162897000MT & productivity of 17.3 MT/ha (NHB 2014-2015). Organic cultivation is the form of agriculture that relies on techniques such as crop rotation, green manure, compost, and biofertilizers which helps to maintain soil productivity. Organic farming excludes or strictly limits the use of chemical fertilizers, pesticides (herbicide, insecticides and fungicides), plant growth regulators, livestock feed additives, antibiotics and genetically modified organisms.

- India ranks 10th among the top ten countries in terms of cultivable land under organic certification (APEDA, 2015).
- The total area under organic certification is 5.71 million Hectare (APEDA: 2015-16). The certified area includes 26% cultivable area with 1.49 million Hectare and rest 74% (4.22 million Hectare) is forest and wild area for collection of minor forest produces [APEDA].

India is primarily agrarian country and this sector provides livelihood to a major part of the population. The indiscriminate use of agro chemicals since green revolution resulted in adverse effect on the soil fertility; crop productivity; quality of produce and more specifically on the environment system. Thus, the importance of organic sources of nutrients was recognized in current scenario in order to get higher yield without disturbing soil health. In this context integrated use of inorganic and organic sources of nutrients in crop production is becoming very crucial for assurance of food security on sustainable basis, which in turn not only improve the soil fertility for sustained crop productivity but also to reduce the cost of inorganic fertilizers.

Vermiwash

Recently, vermiwash production has drawn the attention of commercial vermin-culturists. Vermiwash is a watery extract of compost, the wash of earthworms present in the medium and is honey brown in color.

Principle

Worm worked in the soils have burrows formed by the earthworms. Bacteria richly inhabit these burrows also called as the drilospheres. Water passing through these passages washes the nutrients from these burrows to the roots to be absorbed by the plants. This principle is applied in the preparation of vermiwash.

Vermiwash-what it is?

One of the byproducts of vermin-culture & vermin-composting industry is vermiwash, apart from vermin meal or vermin protein. Vermiwash is the organic fertilizer decoction obtained from units of vermiculture/vermicompost as drainage. It is the liquid extract collected after the passage of water through the different layers of worm culture unit. This decoction contains excretory products of earthworm secretions of the worm-the coelomic fluid oozing through the dorsal pores, mucus, enzymes secreted by the worm & microorganism, plant nutrients, vitamins & plant growth promoting substances which in-creases the resistance power of crops against various diseases and enhances the growth and productivity of crops (Zambare *et al.*, 2008). It is a collection of excretion called vermicasts, which contains lot of nutrients particularly soluble K, Ca & Mg which find their way into vermiwash. As storehouse of nutrients and microorganisms, Vermiwash is used as a foliar spray for crops. Vermiwash is liquid manure that has been reported to have ex-cellent growth promoting effects besides serving as biopesticides even though much work has been done on vermicomposting only few reports are available on vermiwash and its impact on the plant growth (Hatti *et al.,* 2010). The increasing trend of abundant use of inorganic fertilizers along with herbicides and pesticides and exploiting available water resources etc, in the present agriculture system poses a great threat to the sustainability of our agro-ecosystem. Under such situation it is essential to look for alternatives which are effective and eco-friendly.

Chemical composition

P^H	7±0.03
EC	0.25±0.03
Organic carbon (%)	0.008±0.001
Nitrogen (%)	0.01-0.01
Phosphorous (%)	1.70
Potassium (ppm)	26
Sodium (ppm)	8
Calcium (ppm)	3ppm
Copper (ppm)	0.01ppm
Iron (ppm)	0.06ppm
Magnesium (ppm)	160ppm
Manganese (ppm)	0.60ppm
Zinc (ppm)	0.02ppm
Total heterotrophs (Actinomycetes)CFU/ml	1.79×10^3
Nitrosomonas CFU/ml	1.01×10^3
Nitrobacter CFU/ml	1.12×10^3
Total fungi CFU/ml	1.46×10^3

CFU-Colony forming units

Vermiwash contains beneficial microbes which helps in plant growth & prevent infections.

It contains sugars, phenols. Amino acids, hormones and plant growth substances like IAA, GA3,Cytokinin & humic acids *etc.*

Uses:-

- Vermiwash helps to develop resistance to disease & pests in plants.
- It helps to produce good yield & help in initiating flowering in some vegetable crops.
- Vermiwash is a liquid fertilizer & reported that at 20-30% dilution inhibits the mycelial growth of pathogenic fungi.
- Acts as a biopesticide when diluted with 10%cowurine or neem extract or garlic extract.
- Vermiwash is an ecofriendly natural fertilizer prepared from biodegradable organic wastes and is free from chemical inputs.
- It does not have any adverse effect on soil, plant and environment.
- It improves soil aeration, texture and tilth thereby reducing soil compaction.

- It improves water retention capacity of soil because of its high organic matter content.
- It promotes better root growth and nutrient absorption.
- It improves nutrient status of soil-both macro-nutrients and micro-nutrients.
- It can also be added to compost pits to hasten the degradation process.
- Vermiwash is the coelomic fluid extraction, which have enzymes, which stimulate the growth and yield of crops.

Methods of vermiwash production

- ECO-SCIENCE research foundation method.
- Ismails method
- Karunas method
- Economic technique
- KAU's method
- Using plastic drum (1000 lit) method
- Households device
- Kales method
- Earth worms body fluid method.

Ismail method

Ismail (1997) has suggested a method of preparation of vermiwash using a barrel in which a layers of cattle dung and hay were placed on the layer of soil. The epigenic earthworms were introduced to produce compost at faster rate. The vermiwash was collected from the bottom of the barrel after sprinkling water through perforated mud or metal pot from the above.

Karuna's method

In this method, clitellated earthworms were collected from the culture bins. The worms were cultured on organic wastes comprising mostly of leaf-litter, weeds, stubble mixed with one part of cow dung by volume in the form of slurry. Earthworms weighing 1kg were placed into a dry enamel tray for 15-20min to clear out the cast that will be excreted due to handling. Earthworms were then carefully separated from their excreta and then added into glass/ plastic bowl having 500ml of distilled water, at processing temperature of 37°-40° C. The worms were agitated for 3-5min and removed and then added into another bowl containing 500ml of water at room temperature (25°-27° C) to risen thoroughly to collect the exduates adhering to its body wall before releasing back to the culture bins. The contents in the two bowls were mixed to use as a spray. The exduates thus collected was a syrupy, light yellow fluid.

Economical Technique

The method consists of an earthen pot of 8-10kg capacity, which is filled with pieces of stone up to 10cm height from the bottom. Above this, a plastic net is placed and spread out. Layers of coir fiber along with humus containing 1500-2000 earth worms were placed. The bottom of the pot was fixed with a water tap through which vermiwash will be collected. Everyday, the kitchen waste is put into the container. The composting process is allowed to continue for about 10days till brownish black mass of compost is obtained. Regularly, 2-3 tablespoons of fresh cow-dung slurry are poured on the humus as a feed for the multiplying earthworms. The entire mass of compost is soaked with 2liter of water. After 24hrs, about 1.5liter of vermiwash can be collected. The less enriched compost that remains in the pot can be collected and used as fertilizer. Later, the earthen pot can be emptied and set up again to continue the process.

Kerala Agricultural University Method

The method consists of plastic basin having a capacity of 20liters, a plastic perforated waste paper basket and PVC pipe of 5cm diameter and 30cm length. Initially the waste paper basket is covered with a nylon net and placed at the center of the basin upside down. The waste paper basket is covered with nylon net and placed at the center of the basin upside down. A hole is made at the bottom of the waste paper basket so that a PVC pipe of 5cm diameter can be placed in to the basin through the waste paper basket and collects in the PVC pipe, which can be siphoned out by the kerosene pump. Surrounding the waste paper basket, a layer of brick pieces are placed and a layer of coconut fiber of 2-3cm placed above it. The whole apparatus is moistened once and later 2kg of earthworms were introduced and then about 4-5kg kitchen waste is spread over it. Everyday 2liters of water is sprinkled over the compost containing earthworm. After about 7-10 days the kitchen waste turns into dark well decomposed compost material and the vermiwash wash was collected and used. The whole apparatus can be reused by disposing the compost and earthworms.

Plastic Container Method

A plastic container of 65litre capacity was filled with an outlet tap is kept on a stand. Bottom 2-3 cm is filled with pieces of bricks followed by 2 mm sand up to 10cm thickness. Now decomposed cow dung making layer of 15 cm thickness is placed over sand layer. Earthworms numbering 1000 are introduced over cow dung and the remaining space is filled with vegetable waste and litter. It is watered which flows to bottom after washing the worms. The washing is collected in a container placed below the tap fitted at the bottom of the bucket.

Simple Method

A simple method of extraction of vermiwash has been developed capable of yielding about 1.5 litre of vermiwash a day. The system consists of 100 litre plastic barrel, in which a perforated plastic waste paper basket is placed at the centre, upside down, a central 1meter plastic tube 5cms wide with hole upto 7.5cm height at the dipped end. Arrange the waste paper basket, such that the end of the holes touches the base of barrel. The whole unit is kept inside the thatched shed or any cool corner of a farm shed. Above the mixed brick and sand, comes a layer of humus, along with about 5000 earthworms. Everyday, 2-3 table spoons of fresh cow dung slurry is spread over the humus layer, and or vegetable waste, fruit peels, etc as feed for the earthworms. And about 2liter of water is poured from top. The water percolates down and collects at the bottom of the plastic barrel, this water brings along with part of the casts of the earthworms as vermiwash.

Giradd's Method

A method of extraction of earthworm wash, a plant promoter substance was evaluated by Giradd in 2001. In order to have simple devices of collection of earthworm secretions an equipment using GI sheets or plastic crates has been used. The equipment consists of "culturing compartment", joined to collection chamber at the bottom. While water sieve is kept enclosed on the culturing chamber. Both water sieve and culturing compartment have got small holes to allow water from sieve to culturing compartment and from culturing compartment to collection chamber. In another similar equipment , movement of water from culturing compartment to collection chamber is regulated by inserting a Divider in between two chamber which is regulated by inserting a divider in between two chambers. Number of earthworms quantity and quality of raw material used, duration of earthworms culturing and time allowed for water stagnation in the culturing chamber would decide the quality of vermiwash prepared.

Earthworm body fluid method

In certain areas, the farmers ground earthworms and the earthworms paste is mixed in water to be used as foliar spray. 1kg adult earthworms is collected and without any mixing of the cast they are released into a trough containing 500ml of luke warm water (37° -40 °C) and agitated for 2min. Earthworms are now taken out and again washed in another 500ml water at room temperature and released back into the tanks. The agitation in luke warm water makes the earthworms to release sufficient quantity of mucus and body fluid. Transferring into ordinary waste is to wash the mucus sticking to their body surface and this also helps the earthworms to revive from the shock. In this way it is possible to collect only the body fluid without harming the earthworms.

Kale's Method

Kale in 1998 described this method consisted of an outer and inner containers. The inner container has an outlet at the lower side of the vessel. The inner container was filled with decomposing organic matter and about 1-2kg earthworms were accommodated in 12-16 liter capacity vessel. As the earthworms were accommodated in waste, water was slowly added into vessel in excess. The excess water flowing out through the outlet as thick syrup fluid was collected in the outer container. The collected fluid was siphoned out and after diluting was used as foliar spray to different crops. In another method, the cement tank was built at an elevated place from which they want to collect the wash. The slope provided in the tank provided scope for excess water to flow out in drops as thick syrupy emulsion through a small outlet. This was collected in a container and stored in bottles. Before using it was diluted and sprayed to vegetable and other crops.

House hold's method

Developed at Sri Paramakalyani centre for environmental sciences, Alwarkurichi, India. This device consists mainly of a broad based plastic tub with a holding capacity of 10-15liters, a plastic perforated funnel and a hand pump. The funnel is placed upside down with its wide open end covered with a nylon net to allow only the waterly extract and is kept in the center of the plastic tub. A layer of pebbles and sand is placed around the perforated plastic tub for easy percolation of excess water. Three kilograms of vermicompost is spread outside the plastic funnel in the tub. Three liters water is added to the tub over the compost, wetting the entire mass. After 24hrs vermiwash collected in the funnels can be siphoned out by inserting a tube that touches the bottom of the funnel. The process of collection of vermiwash can be repeated by addition of fresh compost into the plastic tub.

Using plastic drum method

A 1000 liters plastic drum, with an outlet tap at the base is used. The idea is to build a living compost for the worms, and pour water through it to collect the vermiwash. The lowest layer is gravel above which biogas slurry, sand, red soil and then biomass half decomposed compost and fresh green leaves are added. This is left as such for two weeks. To remove heat, more compost was added and then also the worms plus the household garbage. This lasts for 3months. 20liters of water is added every second day and 20liters of liquid is collected from the bottom. It can be diluted in the ratio of 1:3 or 1:5 and gives 60liters each time. The vermiwash obtained can be used as a foliar spray for extra nutrient.

Most commonly and commercially followed method was Eco-science research foundation method. It is used for preparing vermiwash in large scale, which is simple, easy and economical for the farmers. Economic technique, simple methods are followed to prepare in small quantity from vegetable waste and is used for kitchen garden for nutritional need.

General Procedure

- Vermiwash units can be set up either in barrels or in buckets or even in small earthen pots. It is the principle that is important.
- The procedure explained here is for setting up of a 250 liter barrel.
- An empty barrel with one side open is taken. On the other side, a hole is made to accommodate the vertical limb of a 'T' jointed tube in a way that about half to one inch of the tube projects into the barrel.
- To one end of the horizontal limb is attached a tap. The other end is kept closed. This serves as an emergency opening to clean the 'T' jointed tube if it gets clogged. The entire unit is set up on a short pedestal made of few bricks to facilitate easy collection of vermiwash.
- Keeping the tap open, a 25 cm layer of broken bricks or pebbles is placed. A 25 cm layer of coarse sand then follows the layer of bricks. Water is then made to flow through these layers to enable the setting up of the basic filter unit. On top of this layer is placed a 30 to 45 cm layer of loamy soil. It is moistened and into this area introduced about 50 numbers each of the surface (epigeic) and sub-surface (anecic) earthworms.
- Cattle dung pats and hay is placed on top of the soil layer and gently moistened. The tap is kept open for the next 15 days. Water is added every day to keep the unit moist.
- On the 16th day, the tap is closed and on top of the unit a metal container or mud pot perforated at the base as a sprinkler is suspended. 5 liters of water (the volume of water taken in this container is one fifth of the size of the main container) is poured into this container and allowed to gradually sprinkle on the barrel overnight. This water percolates through the compost, the burrows of the earthworms and gets collected at the base. The tap of the unit is opened the next day morning and the vermiwash is collected.

The tap is then closed and the suspended pot is refilled with 5 liters of water that evening to be collected again the following morning. Dung pats and hay may be replaced periodically based on need. The entire set up may be emptied and reset between 10 and 12 months of use.

Integrated Vermiwash Unit

Vermiwash can also be prepared from Vermicompost by allowing water to pass through Vermicompost container and collecting it. Vermiwash unit may be established with vermi-composting beds for large scale production.

Precautions

i. The tap should be always kept open to collect the washings.
ii. The unit starts yielding good quality Vermiwash after ten days.
iii. The Vermiwash should be stored in cool dry place.
iv. Water should be poured slowly.
v. Do not mix un-decomposed material while, watering.
vi. Do not add any green material.
vii. Do not allow to compact the contents.

Dosage for Use

a. **Root dip/Stem dip**: The seedlings before transplanting are dipped in Vermiwash solution which is diluted 5 times with water for 15-20 minutes and then transplanted. Similarly the cuttings can also be dipped in the solution.

b. **Foliar spray**: Vermiwash is diluted in water 5 times and sprayed on the foliage of crops. It provides the plant with vital nutrients but also helps to control plant disease.

c. **Soil drench**: Vermiwash is diluted 10 times with water and the soil is drenched with the solution to prevent some of the soil borne pathogens.

Benefits of establishing a vermiwash unit

- Vermicompost-vermiwash units- "A livelihood micro enterprise for rural women".

To farmers

- Less reliance on purchased inputs of nutrients leading to low cost of production.

- Increased soil productivity through improved soil quality.
- Better quantity and quality of crops.
- For landless farmers provides additional source of income generation.

To industries

- Cost effective pollution abatement technology.

To environment

- Waste creates no pollution, as they become valuable raw material for enhancing soil fertility.

To national income

- Boost to rural economy.
- Saving in purchased inputs
- Efficient utilization of wasteland.
- Provides rural employment.

Effect of Vermiwash on Different Crops

Jadhav *et al.*, (2015) studied the effect of vermiwash spray on growth parameters of radish cv. local variety and they concluded that significantly higher leaf length (42.22 cm), maximum leaf weight (16.30 g), higher no. of leaves per plant (11.80), maximum root diameter (5.53 cm) and maximum root length (25.90 cm) was observed with the application of treatment T_4 (1:4) and these were statistically at par with the treatment application of T_3 (1:3) while minimum was recorded with the treatment application of T_5(control).

Jadhav *et al.*, (2015) studied the effect of vermiwash spray on yield parameters of radish cv. local variety and they concluded that significantly higher single radish weight (90.90 g), maximum yield per plant (18.18 kg), maximum yield per hectare (60.59 t/ha) and maximum marketable yield (45.45 t/ha) was observed with the application of treatment T_4 (1:4) and these were statistically at par with the treatment application of T_3 (1:3) while minimum was recorded with the treatment application of T_5.

Ansari and Kumar (2010) studied the effect of vermiwash and vermicompost on plant growth parameters of okra and they concluded that significantly higher plant height (45.83 cm), maximum no. of leaves (13), higher stem circumference (3.10 cm) and maximum marketable yield (69.11 t/ha) was observed with the application of treatment T_6 (vermiwash+vermicompost) and these were statistically at par with the treatment applications of T_5, T_4, T_3 and T_2 while minimum was recorded with the treatment application of T_1 (control).

Ansari and Kumar (2010) studied the effect of vermiwash and vermicompost on biochemical analysis and they concluded that the biochemical analysis was found to be non significant. The maximum Fat (3.52%) and Protein (13%) were recorded in the treatment application of T_6 (vermiwash+vermicompost) while minimum was recorded with the treatment application of T_1 (control).

Singh and Chauhan (2015) studied the effect of different combinations of vermiwash of MSW & Neem plant parts on okra plant pest infestation of *Earias vitella* and they reported that application of VM+NF recorded significantly minimum pod pest infestation (0.04 0.13 and 0.12%) at 30, 45 and 75 days whereas significantly minimum pod pest infestation (1.12%) at 60 days was recorded with the treatment application of VW+NL. The control treatment showed maximum pod pest infestation at 30, 45, 60 and 75 days.

Singh and Chauhan (2015) studied the effect of different combinations of vermiwash of MSW of buffalo dung with aqueous extract of neem plant parts on the germination percent and period (days) of okra seed and they reported that application of VM+NB recorded significantly maximum germination (95%) and maximum germination period (11.78 days) was recorded with the treatment application of VW+NF. The control treatment showed minimum germination % and germination period.

Devasinghe and Kularathna (2016) studied the effect of effective microorganism and vermiwash on growth parameters in lettuce and they reported that control treatment recorded significantly maximum number of leaves (7.25), maximum leaf area (1823.9 m^2), maximum fresh weight of shoots (30.43g) and maximum dry weight of shoots (2.74 g).

Devasinghe and Kularathna (2016) studied the effect of effective microorganisms and vermiwash on physiological parameters in lettuce and they reported that application of 1:20 (V/V)+EW (1:100V/V) recorded significantly maximum photosynthetic rate (5.59), maximum conductance to water (0.30), maximum chlorophyll content (19.94) while minimum transpiration rate (4.48) was recorded with the treatment application of control.

Meghvansi *et al.*, (2012) studied the effect of foliar spray of different concentrations of vermiwash on yield attributes of okra under field conditions and they reported that application of vermiwash 5% recorded significantly maximum total number of fruits per plant (75.75) while maximum average fruit length (15.77 cm) was recorded with the treatment application of vermiwash 10%. The treatment vermiwash 25% showed minimum number of fruits per plant and average fruit length.

Sundararasu (2016) studied the effect of vermiwash growth & yielding pattern of chilli (*Capsicum annum*) and he reported that application of vermiwash 50:50 recorded significantly maximum number of leaves (29, 53 and 104 at 30, 60 and 90 days respectively), maximum plant height (26, 62 and 97 at 30, 60 and 90 days respectively), maximum no. of flowers (14 and 43 at 60 and 90 days respectively), maximum no. of fruits (8 and 31 at 60 and 90 days respectively) and maximum weight of fruits (1517 g at 90 days).

Pradeep and Sharanappa (2014) studied the effect of organic farming practices on growth parameters of onion and they concluded that significantly maximum plant height (42.3 cm), maximum leaves per plant (8.1), maximum leaf diameter (1.46 cm), maximum leaf area index (4.26) and maximum total dry matter (7.59 g) was observed with the application of treatment T_{12} BDLM @100kg N eq/ha+ 3sprays of Panchagavya (3%) and these were statistically at par with the treatment applications of T_{11} EBDLM @100kg N eq/ha+ 3sprays of vermiwash (3%) and T_{10} EBDLM @100kg N eq/ha while minimum was recorded with the treatment application of T_1 BDLM @75kg Neq/ha.

Pradeep and Sharanappa (2014) studied the effect of organic farming practices on yield parameters of onion and they concluded that significantly maximum fresh weight of bulb (143.7g), maximum bulb diameter (6.02 cm), maximum neck diameter (1.42 cm), maximum bulb length (5.36 cm) and maximum bulb yield (42.8 t) was observed with the application of treatment T_{12} BDLM @100kg N eq/ha+ 3sprays of Panchagavya (3%) and these were statistically at par with the treatment applications of T_{11} EBDLM @100kg N eq/ha+3 sprays of vermiwash (3%) and T_{10} EBDLM @100kg N eq/ha while minimum was recorded with the treatment application of T_1 BDLM @75kg Neq/ha.

Pradeep and Sharanappa (2014) studied the effect of organic farming practices on quality of onion and they concluded that significantly maximum ascorbic acid (26.1 mg), maximum TSS (14.4%), maximum reducing sugar (3.98%), maximum non reducing sugar (9.05%) and maximum total sugar (13.03%) was observed with the application of treatment T_{12} BDLM @100kg N eq/ha+ 3sprays of Panchagavya (3%) and these were statistically at par with the treatment applications of T_{11} EBDLM @100kg N eq/ha+ 3sprays of vermiwash (3%) and T_{10} EBDLM @100kg N eq/ha while minimum was recorded with the treatment application of T_1 BDLM @75kg Neq/ha.

Conclusion

The indiscriminate use of agro chemicals since green revolution resulted in adverse effect on the soil fertility; crop productivity; quality of produce and more specifically on the environment system. Among the organic sources vermiwash was found to be one of the best sources for sustainable vegetable production. Vermiwash is an ecofriendly natural fertilizer prepared from biodegradable organic wastes and is free from chemical inputs. It helps to produce good yield & help in initiating flowering in some vegetable crops. Vermiwash is the coelomic fluid extraction, which have enzymes, which stimulate the growth and yield of crops. Vermiwash helps to develop resistance to disease & pests in plants.

References

Ambika, S. and Balakrishnan, K. (2015). Enhancing germination and seedling vigour in cluster by organic priming. *Academic Journal*, 10(8): 298-301.

Aminifard, M. H., Aroiee, H., Azizi, M., Nemati, H . and Hawa, Z.E. Jaafar. (2012). Effect of Humicacid on Antioxidant activities and fruit quality of hot pepper (*Capsicum annum.* L.). *Journal of Herbs, Species & Medicinal Plants*, 18(1):360-369.

Ansari, A. A. and Kumar, S. (2010). Effect of vermiwash and vermicompost on soil parameters and productivity of okra (*Abelmoschus esculentus*) in Guyana, *Pakistan Journal of Agriculture Research,* 23(3-4):137-142.

Devasinghe, U. and Kularathna, L. (2016). Effect of effective microorganisms and vermiwash on yield and quality of lettuce (*Lactuca sativa* L.) in hydroponic cultivation. *International Journal of Chemical, Environmental and Biological Sciences* (IJCEBS), 4(1):78-81.

Jadhav, P. B., Kireeti, A., Patel, D.J., Dekhane, S.S., Patil, N.B. and Patil, S.J. (2015). Utilization of vermiwash spray on growth and yield of radish Cv.local variety. *International Journal of Forestry and Crop Improvement*, 6(1):36-40.

Jandaik, S., Thakur, P and Kumar, V. (2015). Efficacy of cow urine as plant growth Enhancer and Antifungal agent. *Advances in Agriculture*, 1(1):1-7.

Jawale, S. A., More, S.D., Zade, K.K. and Arbad, B.K. (2009). Effect of dung, urine and slurry of different farm animals on yield and quality of spinach. *International Journal of Agricultural Sciences*, 5(2):445-447.

Khanal, A., Shakya, S.M., Shah, S.C. and Sharma, M.D. (2011). Utilization of urine waste to produce quality cauliflower. *The Journal of Agriculture and Environmen*t, 12:91-96.

Kumar, P., Rana, D.K., Singh, V. and Nasseruddin Shah, K.H. (2015). Effect of Humic acid on growth, yield and quality of okra (*Abelmoschus esculentus* L.) Cv. Arka Anamika under subtropical conditions of Garhwal Himalaya. *International Journal for Innovative Research in Science and Technology.* 1(8):151-153.

Meghvansi, M. K., Haneef Khan, M., Gupta. R., Gogoi. H. K. and Singh. L. (2012). Vegetative and yield attributes of okra and naga chilli as affected by foliar sprays of vermiwash on acidic soils. *Journal of Crop Improvement*, 26:520-531.

Pradeep, G. and Sharanappa. (2014). Effect of organic farming practices on growth, yield, quality and economics of onion (*Allium cepa*) in dry zone of Karnataka. *Indian Journal of Agronomy*, 59(2):336-340.

Sajid, M., Abdur Rab., Shah, S.T., Ibadullah Jan., Ibsanul Haq., Haleem, B., Zamin, M., Alam, R. and Zada, H. (2012). Humic acid affects the bulb production of onion cultivars. *African Journal of Microbiology Research.* 6(28):5769-5776.

Salman, S.R., Abou-Hussein, S.D., Abdel-Mawgoud. and El-Nemr, M.A. (2005). Fruit yield and Quality of watermelon as affected by hybrids and humicacid application. *Journal of Applied Sciences Research.* 1(1):51-58.

Sangeetha, M. and Singaram, P. (2007). Effect of lignite humic acid & inorganic fertilizers on growth and yield of onion. *The Asian Journal of Soil Sciences*. 2(1):108-110.

Sharma, R., Shah, S.C., Adhikar, K.R ., Shah, P. and Shrestha, J. (2016). Effects of Cattle urine and FYM on yield of broccoli and soil properties. *Journal of Agriculture Search*, 3(3): 157-160.

Singh, K. and Chauhan, H.K. (2015). Potancy of vermiwash with Neem plant parfts on the infestation of *Earias vitella* (*Fabricius*) and Productivity of Okra (*Abelmoschus esculentus*). *Asian Journal Research Pharmaceutical Science*, 5(1):36-40.

Sundararasu, K. (2016). Effect of vermiwash on growth & yielding pattern of selected vegetable crop chilli(*Capsicum annum*). *International Journal of Advanced Research in Biological Sciences*, 3(9):155-160.

Ugur, A., Demirtas, B., Caglar, S., Zambi, O. and Turkmen, M. (2013). Effect of Humic acid application on yield and quality in green vegetables. *24th International Scientific Expert Conference of Agriculture and Food Industry Proceedings.*

Vennila, C. and Jayanthi, C. (2010). Effect of inorganic nutrients and organic foliar spray on growth and yield of okra. *Indian Journal of Horticulture*, 42(1):94-96.

Section 02
Sustainable Development Techniques

6

Green Corridors

Balaji S. Kulkarni and Aakanksha Mahindrakar

Department of Floriculture and Landscape Architecture, College of Horticulture (UHS Bagalkot),GKVK Post, Bengaluru-560065

Introduction

Earth was a beautiful landscape but man has ruthlessly exploited for his greed specially, in the last century. With rapid industrialization and random urbanization environmental pollution has become a serious problem. Over exploitation of open spaces, ever-increasing number of automobiles and demographic pressure has further aggravated the problem.

According to The Hindu, on December 1, 2015 a blanket of smog surrounded Beijing, China, later this smog covered one tenth of China's area. This situation was recently seen in Delhi, after diwali. It was reported that suspended particulate matter (SPM) content in the air was 14-16 times higher than the permissible limits. So, one of the ways to mitigate this pollution is plating of trees and shrubs as green corridors for abatement of pollution and improvement of environment is an effective way.

Green corridor is also known as Green belts/ Green highways/ wildlife corridors. According to www.cpcbenvis.nic.in green corridor is defined as a thin strip of land that provides sufficient habitat to support wildlife, often within an urban environment especially on highways and railway embankments.

Planting of green corridors has the following advantages:

1. It helps to combat global warming and climate change effects.
2. Green house gases (GHG) can be sequestered using trees.
3. Rare Endangered and Threatened (RET) species can be planted as green corridors and this acts as one of the ways of *ex situ* conservation of the region.

4. It reduces the impacts of all types of pollution like air, dust and noise.
5. It provides shade on glaring hot roads during summers.
6. It reduces the soil erosion at embankment slopes.
7. It reduces the effects of wind and incoming ultra violet (UV) radiation.

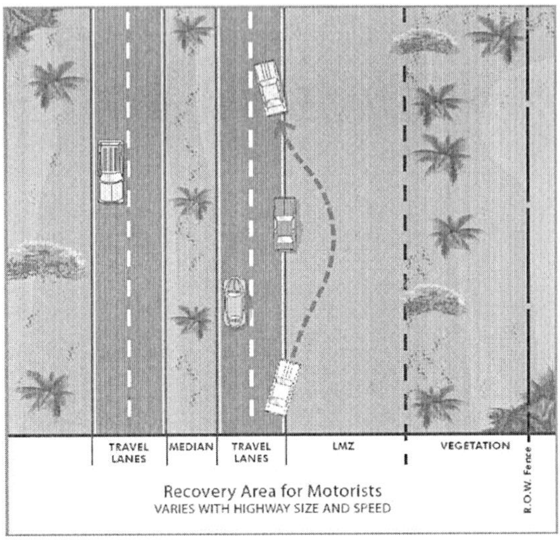

Whenever trees are planted on roadsides, some basic rules have to be followed.

A highway normally has travel lanes, which is separated by median. Median is provided to separate traffic streams and to guide turning movements at intersections. Shrubs are planted on the median; care should be taken to avoid the height of the shrub above 3 feet. Right of way is a narrow strip of land along both sides of a road; it may be a few feet or hundreds of yards wide. This is differentiated into landscape maintenance zone and vegetation zone (Daucus, 2011).

The Landscape Maintenance Zone (LMZ) is the area along the roadsides. It extends from the edge of the road (usually the white line) into surrounding land (right-of-way). The lateral distance or width of the Landscape Maintenance Zone stems from a safety perspective and is intended to create and maintain a recovery area for motorists. The recommended width of the Landscape Maintenance Zone depends on the posted speed as listed in Table 1.

Table 1: Recommended widths of LMZ depend on the posted speed limits

Posted speed (mbh)	Rural Landscape Maintenance Zone recommended distance away from the travel way (ft)
20	8
25	16
30	16
35	16
40	20
45	20
50	30
55	30
60	32

Recommended widths of the Landscape maintenance Zone (LMZ) depended on the posted speed limits but may be changed by the Engineer.

Landscaping at intersections

Rural vegetation near intersections should be mowed or cut within the LMZ, and heights should not exceed 2 ft. This requirement for shorter vegetation should be maintained within certain distances from any given intersection.

Maintain rural vegetation to 3 ft. tall or less within the rural LMZ. In intersections, vegetation 2 ft. tall or less is acceptable. Ensure that vegetation does not create a significant impact hazard or otherwise affect road safety, visibility or hinder drainage systems. Maintain sight distances for passing, signs, stopping and at intersections. A minimum urban horizontal clearance or operational clearance of 18 inches shall be provided between the face of the curb and any obstructions, including trees or shrubs. Vertical clearance for any overhanging vegetation, such as branches coming into the LMZ, should be 17 ft. above the ground level within 2 ft. of the travel way. Vertical clearance above sidewalks should be 7 ft.

Signs

It is important that signs remain visible to road users at adequate distances to ensure safe driving. Vegetation should not impair visibility of any signs: Low stature rural vegetation, less than 3 ft. in height, should be maintained. In curved sections of road, where site distances may be limited, trim vegetation back only according to LMZ requirements. If used, chemical treatments should not extend more than 2 ft. from signs.

Paper 1: Landscape improvement impacts on roadside safety in Texas (Mok *et al.*, 2006)

Objective: To test the effect of landscape improvements on driver performance. This study used a comparison of before-and-after crashes as a quantitative measure of roadside greening.

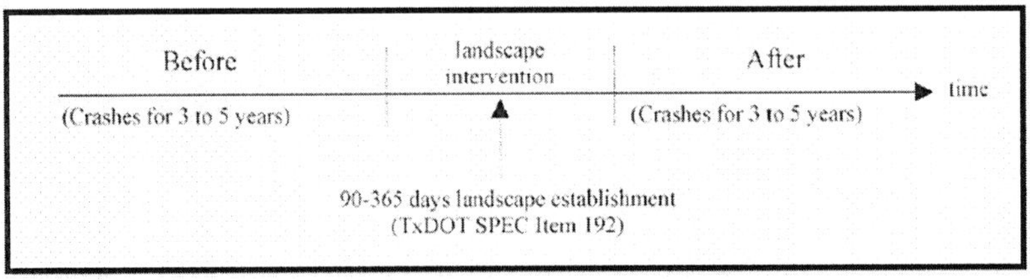

Fig 1 : Concept diagram of data collection before and after landscape improvements

Locations for 3–5 year periods to test for any change of crash rate in sections were studied in 10 locations with landscape treatments. Multiple cases at different times and different locations are selected to enhance statistical validity and minimize the unknown threats and the 'uniqueness' of a particular treatment location. Conceptual explanation of crash data collection and unknown research threats for the comparison period at each study location is delineated in Fig 1. The number of crashes was converted into the crash rate based on the number of crashes per one million Vehicle Miles Traveled (VMT) because in most cases traffic tended to increase over the study period. The calculation formula of crash rate per one million VMT is as follows:

$$\text{Crash rate (CR)} = \frac{\text{Number of crashes} \times 1000000}{\text{Traffic volume} \times \text{Section length} \times \text{Time period (years)} \times 365}$$

Table 2 : Number of crashes and crash rates before and after landscape improvements

Location	Number of crashes			Crash rates		
	Before	After	Difference	Before	After	Difference
Interchange landscaping	1	2	+1	0.0157	0.0564	+0.0407
Roadside landscaping	313	315	+2	2.9545	2.7285	-0.2260
Roadside landscaping	2694	1202	-1492	1.3682	0.5939	-0.7743
Median landscaping	15	11	-4	0.6340	0.3264	-0.3076
Roadside landscaping	139	89	-50	0.4822	0.2447	-0.2375
Interchange landscaping	32	81	+49	0.2749	0.5970	+0.3221
Roadside landscaping	2	2	0	0.0846	0.0778	-0.0098
Roadside landscaping	227	173	-54	0.4296	0.4171	-0.0125
Median landscaping	320	128	-192	1.7231	0.6684	-1.0547
Sidewalk improvement and roadside landscaping	64	64	0	2.2802	2.1005	-0.1797
Total average	3807	2067	-1740	1.0250	0.7811	-0.2439

Eight of 10 study sites showed a decrease in crash rate and two sites showed an increase in crash rate after the landscape treatment. Both of the sites that showed increases in crash rate were complex grade separated interchanges. There were also two outstanding positive sites, study locations nos. 3 and 9. The results show that there was a significant decrease in crash rate after landscape improvements at the 95% confidence level. It could be statistically valid even if the number of sample sites is 10 but the crash rates were calculated by analyzing 5874 crashes based on the control of traffic volume and section length.

Fig 2 : Photos of interchange landscape treatments and planting at location 3 and 9.

Table 3 : Number of tree collisions before and after landscape improvements

Location	Number of tree collisions		
	Before	After	Difference
Interchange landscaping	0	0	0
Roadside landscaping	2	1	-1
Roadside landscaping	18	3	-15
Median landscaping	0	1	+1
Roadside landscaping	1	1	0
Interchange landscaping	0	0	0
Roadside landscaping	0	0	0
Roadside landscaping	1	0	-1
Median landscaping	0	0	0
Sidewalk improvement and Roadside landscaping	2	1	-1
Total average	24	7	-17

The calculated reduction factor of tree collisions shows a decrease of about 70.83% of tree collisions after landscape treatments at ten study locations. As shown in Table 3, there were no extreme changes in tree collisions before and after landscape treatments except for location no. 3.

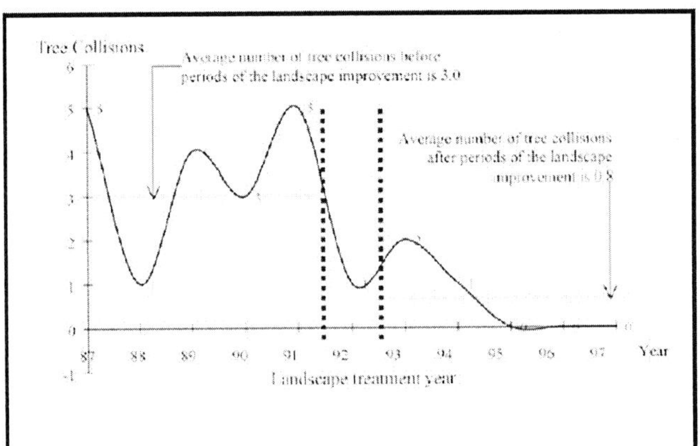

Fig 3 : Number of tree collisions at study location no. 3

From the above figure it is clear that the number of tree collisions in location 3 reduced after landscape intervention. It is recorded that before the landscape interventions tree collisions was recorded to be three and after it was 0.8.

Conclusion

The findings suggest that the use of roadside landscape is having a positive effect on overall performance and can be used as a tool to improve the safety performance of urban streets if the specifics of the affect on driver performance can be identified. There is a correlation between certain types of landscape treatments and reduction in crash rates, indicating that the landscape along the roadside is having a positive affect on driver behavior and perception.

Highways pass through various urban and rural areas joining rural centers with urban markets. The rural areas greatly benefit from the highway construction due to reduction in travel cost and time to reach the market. The industries, which are generally located outside the urban areas, also get benefitted from the highways as the supply of raw materials and delivery of goods become easier and cheaper reducing the overall cost of production. The highways are thus considered to be growth engine for economic development of a country.

Why the need for planting of trees on roadsides?

According to the State of the World's Plants report 2016 by Royal Botanic Gardens, Kew, UK one in five plants are estimated to be threatened with extinction (Bachman, 2016). 2.66% (773) of total species of the world are threatened by transportation and service corridors, 2.35% (684) are threatened by roads and railroads (Royal Botanic Gardens, Kew, UK, 2016).

Highways are acknowledged as the growth engine of economic development. Interstate road infrastructure promotes trade, tourism and bridges geographical divide amongst cities. The fact that it accounts for 2% of our road traffic but contributes 40% of automobile traffic signifies its importance in economy of country.

According to Intergovernmental panel for climate change (2015) transportation sector emits 28% of total greenhouse gases is second to power sector. According to International transport forum (2010) including rails, roads and inlands, out of which road transport contributes 73% of greenhouse gas emissions. Current carbon dioxide emission level is 391 MT and is expected to reach 966 by 2030 (CPCB, 2010).

Urban heat island is a city/ metropolitan city that is significantly warmer than it's surrounding areas due to human activities. The temperature difference is larger at night than during the day and it is most apparent when the winds are weak. Hence, it is important to take proactive measures for pollution containment along the roads.

In the view of this, the Ministry of Road transport and Highways (MoRTH) came up with the National Green Highways policy. The National Highways Authority of India (NHAI) was constituted by an act of Parliament, the National Highways Authority of India Act, 1988. It is responsible for the development, maintenance and management of National Highways entrusted to it. The national green highways policy in 2015 with a vision "to develop environment friendly national highways with participation of community, farmers, NGOs, private sectors, institutions, government agencies and forest department for economic growth and development in a sustainable manner." NGHP came up with National green highway mission with the following objectives:

Assam tribune recently reported the first NHGM was launched in Guwahati on Aug 10, 2016.

National Green Highway Mission Launched Sivasish Thakur

Guwahati, August 10 – The ambitious National Green Highway Mission (NGHM) began its journey today in the North East with the first sapling being planted on the Jorabat-Sonapur section of National Highway-37.

NGHM has laid out an ambitious plan of planting ten crore shade trees along the national highway and the National Highways Authority of India (NHAI) has set aside one per cent from the road construction budget for the project. The project also plans to sequester approximate 12 lakh metric tonnes of carbon from the atmosphere in the whole country, which will have a dramatic impact in cleaning the atmosphere.

Chief General Manager (NE) of NHAI Raj Chakravarty planted the first sapling in presence of senior NHAI officials.

Making an all out effort to enhance the green cover along the tracks, Indian railways firmed up an ambitious proposal to plant 5 crore saplings ahead of onset of monsoon on land owned by it. Only indigenous trees will be planted and railways will also rope in NGO's for carrying out of plantation exercise. The tree variety chosen are mainly of pilkhan and jamun and the saplings will be region specific and vary from zone to zone.

Employment for people

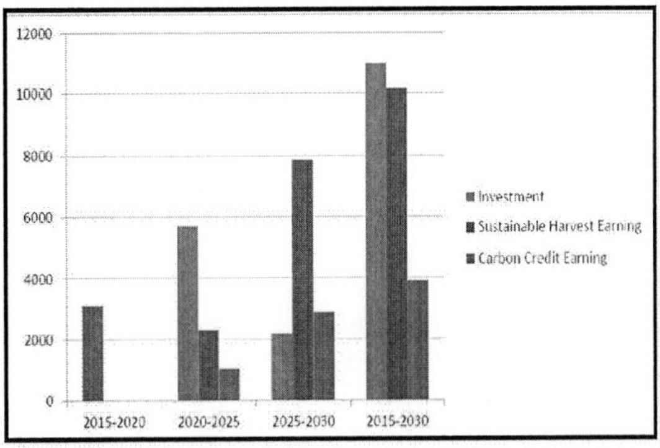

Fig 4: Expected revenue streams in next ten years of project

1. It is expected to sequester 3 MT C/ac/yr, carbon credit rates in voluntary carbon offset markets have been assumed to be 80 dollar per MT during 2025-30.
2. It is expected to earn rs 50,000 acre/ yr from sustainable harvesting practices.
3. It is expected to invest lump sum Rs 12 lakh per km (4.5 acre) in roadside plantation and management for next five years.

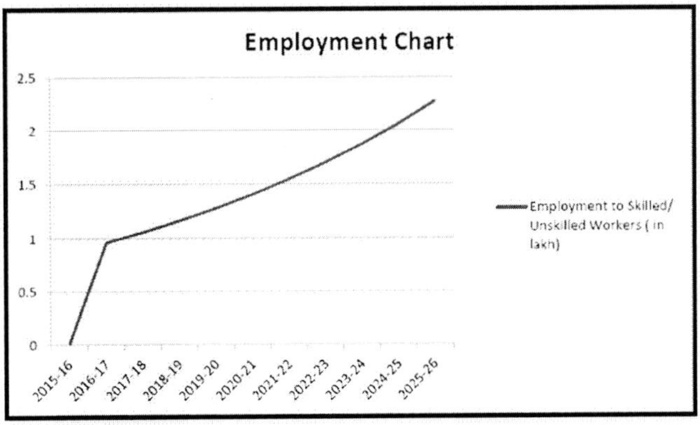

Fig 5 : Employment chart

Plantation and allied activities require large manual workforce, generating livelihoods, oppurtunities for semi/unskilled workers. It is expected to provide 4000[3] man-days of employment per km in plantation and management activities. Plantation and management activities along 1 lakh k stretch will generate 40 crore man cror days of employment ensuring sustained employment oppurtunities for 1 lakh semi/ unskilled workers for minimum 250 days a year for next 10 years.

The four main ways trees combat the ill effects:

Temperature reduction and other microclimatic effects

Removal of air pollutants

Emission of volatile organic compounds and tree maintenance emissions

Energy effects on buildings (Nowak, 2002)

Air Pollution

Number of vehicles has been growing at an average pace of 10.16% per annum (NHAI, 2016). The vehicular pollutants are CO, unburned hydrocarbons, lead, NOx, SOx & suspended particulate matter which emits from tail pipes (Sharma, 2011). Once pollutants are released into the atmosphere, only the plants are hope which can mop up pollutants by adsorbing and metabolizing them from atmosphere. Therefore, plants role in pollution abatement has been increasingly recognized.

Paper 2: Roadside plants as bio indicators of air pollution in an industrial region, Rourkela, India. (Rai and Panda, 2015)

Objective: a. To estimate and analyze air quality of Rourkela and classify it according to Air pollution index (API)

b. To study traits based on biochemical and enzymatic responses of some common roadside plant species growing in selected sites.

The study was conducted in four areas:

Site-A: Station road is an old and densely populated area of the city with high traffic. It is also a commercial area.

Site-B: Plant side road, an industrial area. This area comprises of the Steel Authority of India Limited (SAIL) and experience heavy load of traffic.

Site-C: Sector area is a planned residential area with medium traffic. It is a semi urban area with religious places, recreational areas etc.

Site-D: Hamirpur area is a rural area, mainly agricultural fields, few educational institutes and few colonies. It experiences less traffic load.

$$API = 1/3(SO2/Sso_2 + NOX/S_{NOX} + SPM/S_{SPM}) \times 100$$

Air pollution index as calculated by CPCB, New Delhi, where S_{so2}, S_{nox}, S_{SPM} denotes permissible limits.

Table 4 : Ambient Air Quality and Air Pollution Index for different sites during the study periods

Site	RSPM (μg m^{-3})	SPM (μg m^{-3})	NO_2(μg m^{-3})	SO_2 (μg m^{-3})	API
Site-A	190.8	538.32	20.41	17.65	72.75 (MAP)
Site-B	220.12	540.09	23.62	21.65	79.50(HAP)
Site-C	165.92	410.28	18.88	11.77	59.00 (MAP)
Site-D	128.74	370.04	11.98	9.89	48.50 (LAP)

Where, API index values are categorized into:

Index value	Remarks
0-25	Clean air
26-50	Light air pollution
51-75	Moderate air pollution
76-100	Heavy air pollution
>100	Severe air pollution

Ambient air quality monitoring at different sites of Rourkela city indicated that Site B (industrial area) was highly charged with pollutants emission from steel plant (SAIL) and automobile exhaust and Site D (rural area) showed minimum level of air pollutants when compared to other two sites. Air quality index for the selected sites in Rourkela city is shown in Table 4. On the basis of air pollution index, site B was categorized as heavy air pollution site with air pollution index 79.50 , site A and site C as moderately air pollution site with air pollution index 59.00 and 72.75 respectively and site D as light air pollution site with air pollution index 48.50 (Table 4) . The values of all the four air pollutants were lowest at site D because it is a rural area with less number of vehicles and highest at site B as it is an industrial area with some commercial complexes and ample no of heavy vehicle passes throughout the day due to the presence of Steel Authority of India Limited (SAIL).

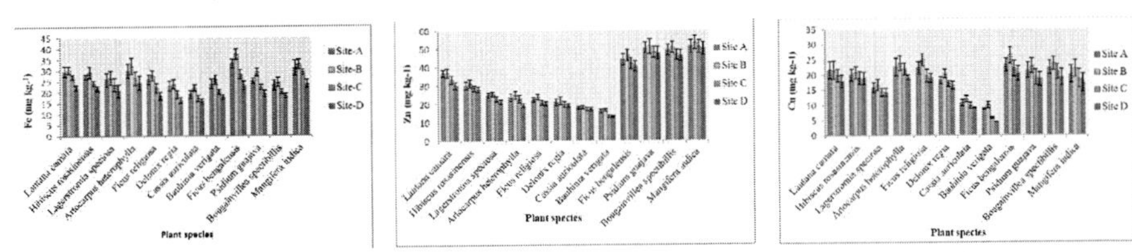

Heavy metal concentrations in the leaves of studied sites are shown in figure 2 (A, B and C).Heavy metal concentration were ranked in the order of Zinc >Iron > Copper in the studied species. Among the examined plant species, *Ficus bengalensis* and *Psidium guajava* showed the highest heavy metal concentration while the lowest were observed in *Cassia auriculata* and *Bauhinia variegate.* The elevated emission of heavy metals concentration in our study sites can be attributed to industrial emission and a greater density of heavy vehicles. However, heavy metal accumulation in plants did not follow any particular pattern, which might be due to their inherent metal accumulation capacity, Variation in growth rate and stage of maturity (*Gupta et al.* 2012; Pandey *et al.* 2009; Sharma *et al.* 2008). The maximum concentration of Fe, Zn and Cu in the airborne dust particulates of site B might be due to the presence of Steel Authority of India Limited (SAIL) and heavy traffic density.

Table 5 : Mean concentration of total chlorophyll of selected plant species at different sites.

Plant species	Site A	Site B	Site C	Site D
Ficus bengalensis	0.45±0.06	0.42±0.06	1.10±0.04	1.46±0.03
Ficus reliogosa	0.32±0.04	0.30±0.03	0.82±0.03	1.08±0.04
Mangifera indica	0.49±0.07	0.46±0.07	1.12±0.03	1.84±0.04
Bougainvillea spectabilis	1.87±0.03	1.78±0.06	2.07±0.03	2.02±0.04
Psidium guajava	1.12±0.04	1.05±0.07	1.54±0.04	1.89±0.08
Hibiscus rosa-sinensis	1.43±0.05	1.36±0.06	1.93±0.04	2.10±0.11
Lantana camara	1.59±0.04	1.11±0.02	1.82±0.12	2.00±0.04
Delonix regia	0.89±0.13	0.75±0.02	0.91±0.04	1.08±0.12
Artocarpus heterophullus	0.88±0.08	0.83±0.4	0.99±0.04	1.19±0.02
Cassia auriculata	1.58±0.09	1.13±0.04	1.90±0.02	1.99±0.02
Bahunia variegata	0.83±0.04	0.65±0.03	0.86±0.16	1.18±0.05
Lagestroemeria speciosa	1.11±0.04	1.01±0.15	1.23±0.05	1.76±0.02

It was noticed that chlorophyll content decreased with increasing pollution load with maximum reduction at site B and site A that harbours industrial set up and highest vehicular density. The highest total chlorophyll content (2.10±0.04) was recorded in *Hibiscus rosasinensis* at Site D and the lowest (0.30±0.03) in *Ficus religiosa* at Site B. Chlorophyll is said to be an index of productivity and it plays an important role in plant metabolism, hence any alteration in chlorophyll concentration may change the morphological, physiological and biochemical behaviour of the plant.

Table 6 : Mean concentration of ascorbic acid content of selected plant species at different sites.

Plant species	Site A	Site B	Site C	Site D
Ficus bengalensis	0.44±0.04	0.57±0.01	0.35±0.04	0.20±0.01
Ficus reliogosa	0.65±0.03	0.71±0.03	0.22±0.02	0.35±0.03
Mangifera indica	0.53±0.08	0.60±0.02	0.50±0.03	0.43±0.04
Bougainvillea spectabilis	0.23±0.03	0.27±0.06	0.20±0.03	0.15±0.04
Psidium guajava	0.41±0.04	0.50±0.07	0.36±0.04	0.38±0.08
Hibiscus rosa-sinensis	0.45±0.03	0.51±0.06	0.35±0.04	0.32±0.11
Lantana camara	0.33±0.01	0.36±0.02	0.31±0.12	0.25±0.04
Delonix regia	0.56±0.03	0.59±0.02	0.40±0.04	0.26±0.12
Artocarpus heterophullus	0.77±0.03	1.03±0.4	0.54±0.04	0.25±0.02
Cassia auriculata	0.62±0.04	0.70±0.04	0.58±0.02	0.33±0.02
Bahunia variegata	0.72±0.05	0.83±0.03	0.66±0.16	0.35±0.05
Lagestroemeria speciosa	0.62±0.04	0.66±0.15	0.50±0.05	0.20±0.02

The concentration of ascorbic acid at Site B ranges between 0.27±0.04 to 1.03±0.03 and at Site A it ranges between 0.23±0.03 to 0.77±0.03 with *Bougainvillea spectabilis* and *Artocarpus heterophyllus* recording the lowest and highest value. Ascorbic acid is a natural antioxidant and a strong reducing agent. It plays an important role in pollution tolerance and protects the plant against oxidative damage by maintaining the stability of cell membrane during pollution stress and scavenges cytotoxic free radicals. Present study showed elevation in the concentration of ascorbic acid at Site B in all the selected plant species. Pollution load dependent increase in ascorbic acid content of all the plant species may correspond with oxidative stress due to greater accumulation of particulate heavy metals or increased rate of production of reactive oxygen species (ROS) during photo-oxidation process.

Table 7 : Mean concentration of total soluble sugar content of selected plant species at different sites.

Plant species	Site A	Site B	Site C	Site D
Ficus bengalensis	0.54±0.04	0.51±0.01	0.55±0.04	0.56±0.01
Ficus reliogosa	0.21±0.03	0.19±0.03	0.49±0.02	0.54±0.03
Mangifera indica	0.41±0.08	0.38±0.02	1.17±0.03	1.20±0.04
Bougainvillea spectabilis	0.13±0.03	0.12±0.06	0.30±0.03	0.34±0.04
Psidium guajava	0.37±0.04	0.15±0.07	0.56±0.04	0.58±0.08
Hibiscus rosa-sinensis	0.11±0.03	0.10±0.06	0.14±0.04	0.19±0.11
Lantana camara	0.13±0.01	0.11±0.02	0.16±0.12	0.21±0.04
Delonix regia	0.31±0.03	0.28±0.02	0.99±0.04	1.02±0.12
Artocarpus heterophullus	0.54±0.03	0.48±0.4	0.75±0.04	0.77±0.02
Cassia auriculata	0.16±0.04	0.11±0.04	0.60±0.02	0.61±0.02
Bahunia variegata	0.12±0.05	0.10±0.03	0.21±0.16	0.23±0.05
Lagestroemeria speciosa	0.17±0.04	0.14±0.15	0.24±0.05	0.28±0.02

The concentrations of total sugar content were markedly decreased with increasing pollution load at Site B (heavy pollution site) for all the plant species when compared with other three sites and the maximum reduction was seen in *Mangifera indica* (1.20±0.04 to 0.38±0.02 mg/g) and *Delonix regia* (1.02±0.05 to 0.28±0.06 mg/g) Soluble sugar is an important constituent and source of energy for all living beings. Plants manufacture this organic substance during photosynthesis and breakdown during respiration (Seyyednejad *et al.* 2011; Tripathi and Gautam 2007). Reduction in soluble sugar content in polluted site may correspond with a lower photosynthetic rate and higher energy requirements due to airborne heavy metal stress

Table 8 : Mean concentration of total protein content of selected plant species at different sites.

Plant species	Site A	Site B	Site C	Site D
Ficus bengalensis	0.73±0.04	0.69±0.01	0.84±0.04	0.83±0.01
Ficus reliogosa	0.28±0.03	0.21±0.03	0.90±0.02	0.92±0.03
Mangifera indica	0.65±0.08	0.31±0.02	0.83±0.03	0.87±0.04
Bougainvillea spectabilis	0.69±0.03	0.57±0.06	1.21±0.03	0.33±0.04
Psidium guajava	0.77±0.04	0.63±0.07	1.40±0.04	1.31±0.08
Hibiscus rosa-sinensis	0.71±0.03	0.65±0.06	0.80±0.04	0.73±0.11
Lantana camara	0.78±0.01	0.76±0.02	1.18±0.12	1.12±0.04
Delonix regia	0.37±0.03	0.28±0.02	0.66±0.04	0.70±0.12
Artocarpus heterophullus	0.52±0.03	0.49±0.04	0.83±0.04	0.80±0.02
Cassia auriculata	0.76±0.04	0.73±0.04	0.94±0.02	1.06±0.02
Bahunia variegata	0.50±0.05	0.47±0.03	0.88±0.16	0.77±0.05
Lagestroemeria speciosa	0.88±0.04	0.80±0.15	0.99±0.05	0.98±0.02

Maximum reduction was observed in *Ficus religiosa* from (0.92±0.04 to 0.21±0.03 mg/g) and *Bougainvillea spectabilis* (1.33±0.06 to 0.57±0.04 mg/g). Reduction in protein content of plants at Site A and B may be due to the enhanced rate of protein denaturation and break down of existing protein to amino acid or reduced *denovo synthesis* of protein

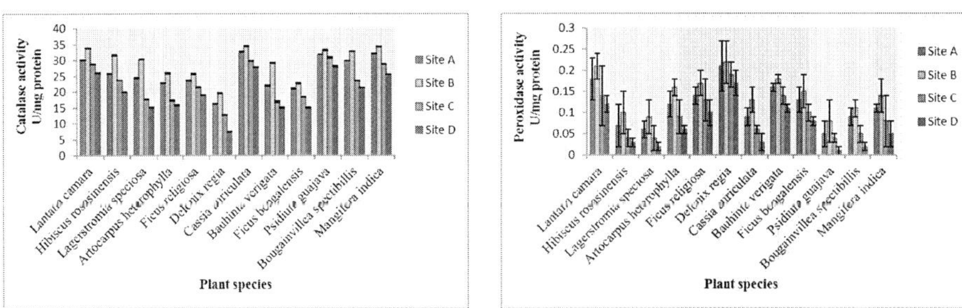

Fig. 7: Catalase and peroxidase activity in leaf samples

The average activity of catalase and peroxidase enzymes were increased in all the plant species at Site B and A (heavy pollution site) compared to Site D and C (Light and medium air pollution site). *Cassia auriculata* (34.40±0.12 U/mg protein) showed highest catalase concentration and *Delonix regia* (7.34±0.08 U/mg protein) showed lowest Catalase concentration. Highest peroxidase concentration was recorded in *Delonix regia* (0.22±0.09U/mg protein) and lowest in *Psidium guajava* (0.01±0.01U/mg protein). In plant cells, electron may be transferred via chloroplast or mitochondrial electron transfer system.

Table 9 : A correlation between ambient air quality and foliar parameters of selected plant species.

Total chlorophyll	Ascorbic acid	Protein	Sugar	
SPM	-0.9706	0.9562	-0.9716	-0.9895
RSPM	-0.9873	0.9948	-0.9136	-0.09168
SO_2	-0.9596	0.9589	-0.9899	-0.9796
NO_2	-0.9602	0.9700	-0.8208	-0.8371

A significant negative correlation was found between air quality parameters and different foliar parameters except ascorbic acid which showed significant positive correlation with pollution load.

Conclusion

The morphological and biochemical traits of selected roadside plant species can serve as suitable bio indicators of particulate pollution and an excellent quantitative and qualitative index of pollution level by capturing significant amount of health- damaging particles from the atmosphere with the potential to perk up local air quality.

Paper 3 : To control the effect of air pollution using roadside trees (Mate and Deshmukh, 2015)

Objective: To analyze the effects of air pollution on tree leaves using APTI method

Table 10: APTI values of trees selected from rourkela (industrial) and Aizawal (Non-industrial area)

Plant species	APTI Value	
	Rourkela	*Aizawal*
Ficus bengalensis	8.64	6.30
Psidium guajava	9.00	6.00
Bougainvillea spectabilis	7.72	6.94
Mangifera indica	8.98	7.95
Lantana camera	8.11	5.95
Hibiscus rosa-sinensis	9.35	7.43

Table 10 shows APTI values of trees which is from Rourkela (Industrial area) and Aizawl (Non-industrial area). In Industrial area *Hibiscus rosa-sinensis* is more tolerant to air pollution and *Bougainvillea spectabilis* is more sensitive to air pollution. In Non-industrial area

Mangifera indica is more tolerant to air pollution and *Lantana camara* is more sensitive to air pollution. Also it states that Rourkela area is more polluted than Aizawl area.

Table 11: APTI value of trees in different seasons

Plant species	APTI Value		
	Monsoon	Winter	Summer
Azadiracta indica	9.45-25.26 **18.00**	12.91-25.94 **18.71**	10.30-26.54 **14.73**
Polyalthia longifolia	11.09-19.47 **14.74**	8.94-28.8 **19.04**	7.98-10 **16.49**
Ficus bengalensis	11.1-17.89 **14.96**	10.15-19.88 **16.36**	11.61-23.98 **16.43**
Mangifera indica	15.81-16.91 **16.41**	14.54-18.43 **15.68**	15.10-18.91 **16.54**
Acacia arabica	10.94-13.21 **11.40**	14.77-19.97 **18.09**	13.63-19.64 **16.46**
Peltophorum pterocarpum	13.15-17.04 **13.75**	8.84-16.75 **16.75**	15.81-19.65 **16.57**

Bold values represent average APTI of all sampling sites

The effects of air pollution on trees which are vary according to seasons.The result shows that according to average value of APTI *Azadirachta indica* is more tolerant and *Acacia arabica* is more sensitive to air pollution in monsoon season, *Polyalthia longifolia* is more tolerant and

Mangifera indica is more sensitive to air pollution in winter season, *Peltophorum pterocarpum* is more tolerant and *Azadirachta indica* is more sensitive to air pollution in summer season.

Conclusion

It is concluded that this study is most important, useful for the better understanding, controlling and management of air quality. It is also useful for selection of suitable plant species for plantation in highly polluted industrial areas, road side areas to reduce the air pollution. The trees having higher APTI value is more tolerant to air pollution and the trees having lower APTI value is more sensitive to air pollution. The tress having a high APTI score has a low acidic pH in their leaves with a high chlorophyll content and maximum relative water content.

Paper 4: Assessment of APTI of some trees in Moradabad city, India (Tripathi *et al.,* 2011)

Objective: To evaluate the APTI of different plant species growing near residential, commercial and industrial area of the city as to provide a particular plant species to grow in respective areas.

Table 12 : Air pollution status of respective sites

	PTC (SI)			Town hall (SII)			Mugalpura (SIII)		
	SPM	SO_2	NO_2	SPM	SO_2	NO_2	SPM	SO_2	NO_2
Mar	174.65 ± 6.15^{1}	15.18 ± 0.73^{1}	17.28 ± 0.71^{2}	279.2 ± 13.9^{4}	42.5 ± 2.07^{1}	41.42 ± 1.89^{2}	854.06 ± 27.78^{1}	5.4 ± 3.20^{2}	60.4 ± 2.20^{2}
Apr	189 ± 7.19^{ns}	15.24 ± 0.37^{ns}	18.19 ± 0.83^{ns}	280.3 ± 13.6^{ns}	42.63 ± 3.17^{ns}	42.53 ± 1.70^{ns}	768.1 ± 31.64^{1}	89.4 ± 2.22^{ns}	61.2 ± 1.98^{ns}
May	191.53 ± 8.32^{ns}	16.87 ± 0.82^{ns}	20.58 ± 0.98^{ns}	282.3 ± 11.9^{ns}	43.8 ± 4.71^{ns}	42.5 ± 2.13^{ns}	855.22 ± 29.50^{1}	87.4 ± 3.95^{ns}	63.4 ± 3.20^{ns}
Jun	214.56 ± 9.65^{ns}	19.64 ± 0.59^{1}	22.19 ± 1.06^{ns}	283.1 ± 12.3^{ns}	45.27 ± 5.1^{ns}	43.6 ± 120^{ns}	863.57 ± 41.09^{ns}	84.1 ± 2.04^{ns}	65.14 ± 4.19^{4}
Jul	126.89 ± 6.20^{s}	11.45 ± 0.48^{s}	8.56 ± 0.41^{3}	174.5 ± 9.79^{3}	22.58 ± 2.2^{s}	33.57 ± 4.20^{2}	496.2 ± 29.19^{3}	74.4 ± 3.36^{1}	57.8 ± 1.91^{1}
Aug	128.21 ± 6.41^{ns}	12.48 ± 0.26^{ns}	9.76 ± 0.40^{ns}	175.87 ± 10.79^{ns}	22.98 ± 1.97^{ns}	34.87 ± 3.98^{ns}	502.5 ± 19.21^{ns}	75.14 ± 4.16^{ns}	59.4 ± 1.78^{ns}
Sep	129.42 ± 3.71^{ns}	14.63 ± 0.32^{ns}	11.58 ± 0.56^{ns}	177.4 ± 11.9^{ns}	24.67 ± 2.19^{ns}	36.38 ± 1.23^{ns}	538.1 ± 21.21^{ns}	77.5 ± 2.32^{ns}	62.74 ± 2.17^{ns}
Oct	130.47 ± 6.20^{ns}	15.67 ± 0.65^{ns}	12.87 ± 0.63^{ns}	178.3 ± 13.3^{ns}	24.4 ± 3.30^{ns}	37.85 ± 0.90^{ns}	558.2 ± 16.36^{ns}	79.1 ± 3.19^{ns}	63.9 ± 2.65^{ns}
Nov	141.52 ± 9.32^{ns}	17.58 ± 0.60^{ns}	20.45 ± 0.25^{3}	243.8 ± 15.8^{3}	65.2 ± 4.37^{3}	60.57 ± 2.96^{3}	589.2 ± 32.13^{ns}	95.1 ± 3.30^{3}	71.8 ± 1.32^{1}
Dec	149.57 ± 5.69^{s}	18.49 ± 0.69^{ns}	21.87 ± 1.53^{ns}	244.8 ± 13.1^{ns}	66.3 ± 3.47^{ns}	61.98 ± 3.59^{ns}	624.3 ± 20^{ns}	97.1 ± 4.20^{ns}	73.1 ± 2.59^{ns}
Jan	162.43 ± 6.90^{ns}	20.48 ± 1.00^{ns}	23.57 ± 1.12^{ns}	247.8 ± 18.69^{ns}	68.17 ± 4.12^{ns}	63.52 ± 2.69^{ns}	671.2 ± 19.13^{ns}	98.4 ± 3.12^{ns}	75.1 ± 1.32^{ns}
Feb	175.42 ± 9.12^{ns}	22.69 ± 0.99^{1}	25.24 ± 1.23^{ns}	248.6 ± 20.6^{ns}	69.28 ± 5.13^{ns}	65.42 ± 2.47^{ns}	580.6 ± 22.04^{1}	101.5 ± 3.30^{ns}	76.2 ± 1.45^{3}
A.A	159.49	16.72	17.68	234.17	44.84	47.03	658.44	87.04	65.84
NAAQS	140	60	60	140	60	60	360	60	80

TAnnual average concentration of major air pollutants i.e. SPM, SO_2 and NO_2 at different locations and National Ambient Air Quality Standard are summarized in Table 12. In the light of values of SPM (658.44 µg m-3), SO_2 (119.83 µg m-3) and NO_2 (98.41µg m-3) it was observed that Mughalpura (SIII) is the most polluted site and PTC (SI), having the concentration of SPM (159.47 µg m-3), SO_2 (61.91 µg m-3) and NO_2 (74.83 µg m-3), is the least polluted area of Moradabad. In Table 12 means of individuals parameters are compared in different months within column.

Table 13 : pH, relative moisture, chlorophyll,ascorbic acid content of the trees growing in different areas

	Plant species									
	Ficus rumphii	*Pongamia pinnata*	*Alstonia scholaris*	*Holoptelea integrifolia*	*Saraca indica*	*Pithecolobium dulcis*	*Cassia simea*	*Bauhinia variegata*	*Azadiracta indica*	*Grewelia robusta*
Residential area (SI)										
pH	12.25 ± 1.44^{3}	5.76 ± 1.60^{3}	11.2 ± 1.41^{2}	9.41 ± 1.06^{ns}	9.1 ± 0.88^{ns}	10.53 ± 1.20^{3}	4.22 ± 0.90^{3}	8.48 ± 0.93^{3}	9.15 ± 1.15^{3}	12.5 ± 1.66^{3}
Rh	68.12 ± 1.89^{3}	40.81 ± 1.72^{ns}		72.4 ± 1.01^{3}	58.5 ± 1.05^{3}	50.41 ± 1.50^{3}	70.58 ± 1.43^{3}	42.26 ± 1.46^{3}	55.31 ± 1.33^{3}	
70.62 ± 0.54^{3}	65.8 ± 1.51^{2}									
Total chlorophyll (mg g^{-1})	12.1 ± 1.09^{1}	8.31 ± 0.88^{1}	11.2 ± 1.41^{ns}	12.32 ± 1.96^{ns}	10.2 ± 1.11^{ns}	9.7 ± 0.92^{ns}	4.3 ± 0.91^{2}	9.38 ± 1.49^{2}	12.7 ± 1.44^{ns}	11.7 ± 0.99^{ns}
Ascorbic	10.62 ± 1.49^{3}	2.64 ± 1.15^{3}	8.6 ± 0.94^{3}	10.21 ± 0.85^{ns}	10.5 ± 1.33^{ns}	9.15 ± 1.08^{ns}	2.65 ± 0.53^{s}	11.46 ± 0.85^{3}	7.65 ± 0.52^{2}	

According to observed APTI values *Ficus rumphii* shows the highest value (32.56) at PTC which is a residential site and *Holoptelea integrifolia* shows the highest value at Townhall site (49.5) as well as Mughalpura site (55.8) which are commercial and industrial sites respectively but *Cassia simea* shows lowest values at SI (6.48), at SII (5.0) and at SIII (6.09).

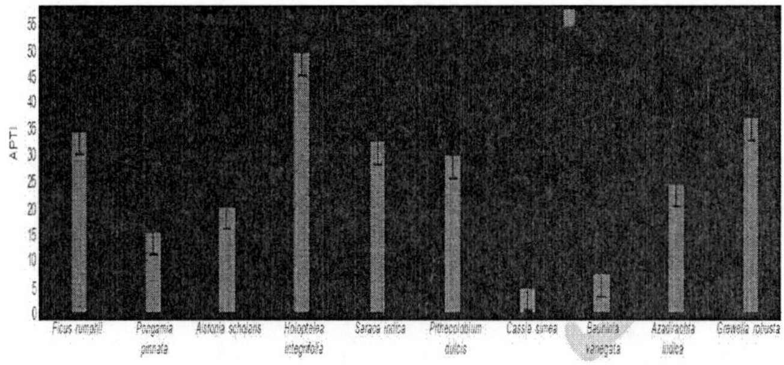

Fig 8 : APTI of trees growing in different sites of Moradabad city ± SD

Although all the species showed significant ($p<0.01$) variation for all the biochemical parameters i.e. total chlorophyll, ascorbic acid, pH and relative water content. The extent up to which plant species were affected varied from species to species and site to site. The results indicate that leaves of all the plant species were healthy and green at SI in comparision to SII and SIII. Among all the ten species only three plant species *Alstonia scholaris, Bauhinia variegata,*

Cassia simea are selected as sensitive species because these species have more APTI values at SI in comparison to SII and SIII.

Conclusion

Green belt plantation around the air polluting unit can never be a claim for the removal of air pollutants at the region, but effectively planted trees in the green belt may potentially remove the toxic gases in considerable amount. So this study is useful for the better understanding and management of air quality as well as in selection of suitable plant species (with high APTI) for plantation in industrial area as well as roadside and this may become one of the strategy for the abatement of city's air pollution because it will have a marked effect on many aspects of the quality of the urban environment and the richness of life in a city.

Noise Pollution

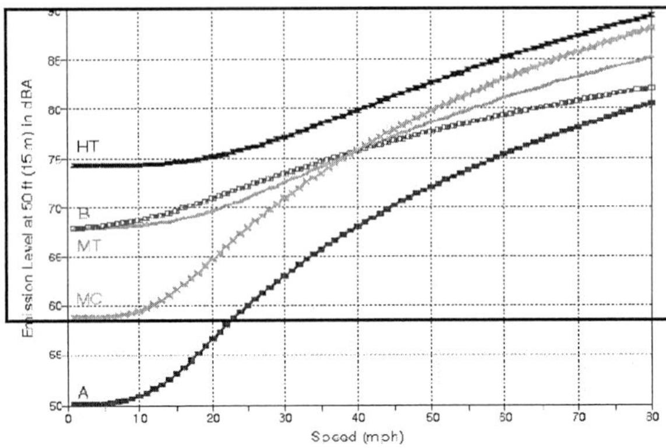

Fig. 9 : Noise caused by different automobiles

In the above figure it can be observed that noise level of autorickshaw, motor cycle, mid truck, bus and heavy truck have sound range in 75 to 90 decible, but according to below table, the normal range varies from 30-65dB(A).

Decible (dBA)	Effect
30-65	Normal
60-95	Psychological disorders
90-120	Hearing and psychological problem
>120	Negative effect on human and all living animals

So, planting "noise buffers" composed of trees and shrubs can reduce noise by five to ten decibels for every 30m width of woodland, especially sharp tones, and this reduces noise to the human ear by approximately 50%. (www.forestry. gov.uk)

Paper 5: Importance of green belts to reduce noise pollution and determination of roadside noise reduction effectiveness of bushes in Konya, Turkey (Onder and Kockbecker, 2012).

Table 14 : Noise measurements and noise reductio2n value of plants

Ist group	Planting species Measuring distances (7m)	Measurement dB(A)	Decreasingd B(A)
1st measure	No vegetation (control area)	72	
2nd measure	*Berberis thunberjia- Cotoneaster dammeri*	67.2	4.8
3rd measure	*Berberis thunberjia- Euonymus japonica*	68.3	3.7
4th measure	*Berberis thunberjia-Pyracantha coccinea*	65.7	6.3
5th measure	*Forsythia intermedia- Pyracantha coccinea- Juniperus horizantalis*	66	6

IInd group	Planting species Measuring distances (11m)	Measurement dB(A)	Decreasingd B(A)
1st measure	No vegetation (control area)	65.1	
2nd measure	*Spiraea vanhouetti- Cotoneaster dammeri*	62.4	2.7
3rd measure	*Spiraea vanhouetti- Cotoneaster dammeri*	56.6	5.5
	Pyracantha coccinea-		
4th measure	*Forsythia intermedia- Juniperus horizantalis*	60.2	4.9

IIIrd group	Planting species Measuring distances (22m)	Measurement dB(A)	Decreasingd B(A)
1st measure	No vegetation (control area)	59.8	
2nd measure	*Pyracantha coccinea-Juniperus horizantalis- Cotoneaster horizantalis*	56.3	3.5
3rd measure	*Pyracantha coccinea-Juniperus horizantalis*	55.9	3.9
4th measure	*Cotoneaster dammeri- Juniperus horizantalis- Spiraea vanhouetti*	53.6	6.2
5th measure	*Pyracantha coccinea- Cotoneaster dammeri*	54.7	5.1
6th measure	*Pyracantha coccinea- Forsythia coccinea- Tamarix tetranda*	54.9	4.9

The plant groups which were consisted from *Pyracantha coccinea* M., *Cotoneaster horizontalis* Decne., *Berberis thunbergii* D.C., *Cotoneaster dammeri* C.K., *Forsythia intermedia* Zab., *Juniperus horizontalis* L., *Spirea vanhouetti* Briot., *Tamarix tetranda* L., *Euonymus japonica* L. were used in a width of 5m, 9m and 20m to determination of decreasing effect on the noise in Konya.

Conclusion

The principles which are needed to be successful in establish of a noise belt area could be summarize such as in the following:

The planting area should be a total of 5m width. This amount could be as much as far up to 30m.

-The plants which are subjected to use should be selected from natural flora or the appropriate varieties which are compatible with the natural flora should be used

-The "evergreen" plants should be used primarily.

-The plants should be planted uprightly to the noise direction.

-The plants should planted closely as possible as to each other.

Paper 6 : The effect of tree species on reduction in rate of noise pollution at the edge og hycarian forest roads (Nasiri *et al.*, 2015).

Objective: To evaluate the impact of broad-leaved and coniferous species on reducing noise pollution according to the distance from the sound source and the tree species type.

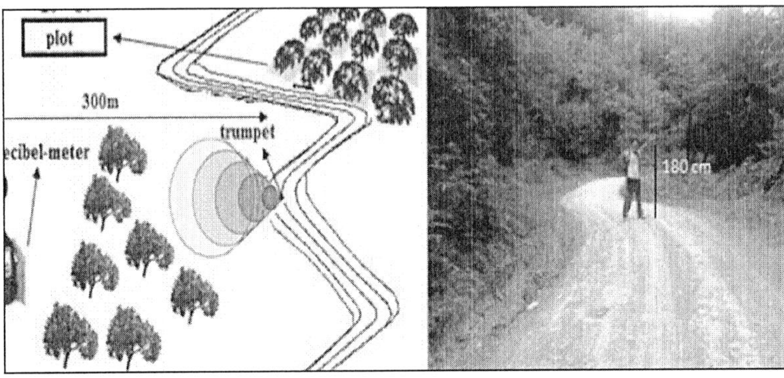

Fig 10 : Measurement method of sound level at the edge of the forest road

A person who located on road center in front of the plot produced noise using trumpet to compare the effects of different species on sound level. Another person who located at the edge of the plot and at the distance of 20 meters from the road measured the sound level using decibel-meter. According to Fig 10 we can find that the weather was cloudy (without rain and wind) on the sound level measurement day and the measurements were implemented at height 180 cm from the ground.

Table 15 : Statistical test for frequency of trees impact rate of sound level

Frequency of trees	10-20	20-30	>30
Sound level (db)			
Hardwood	90.6[a]	84.4[b]	74.0[c]

Frequency of trees	10-20	20-40	>40
Sound level (db)			
Coniferous	91.6[a]	86.6[a]	70.4[b]

a* significantly different at 0.05 probability level

Results showed, the rate of sound level decreased with increasing frequency of broadleaved and coniferous trees. So that, there was significant difference among all classes for broadleaved species, but for coniferous species the significant difference was only observed for class with more than 40 trees per plot (P>0.05). There was also an inverse relationship between frequency of broadleaved and coniferous trees and the rate of sound level.

Table 16 : Statistical test for distance impact on the rate of sound level

Distance (m)	20	100	300
Sound level (db)			
Hardwood	81.00[a]	76.80[a]	72.30[b]
Coniferous	81.40[n.s]	77.60[n.s]	75.20[n.s]

a* significantly different at 0.05 probability level; n.s- Non significant

There was a significant difference between the distance of 300 meter as compared with the distances of 20 meter and 100 meter for the broadleaved species ($P<0.05$).

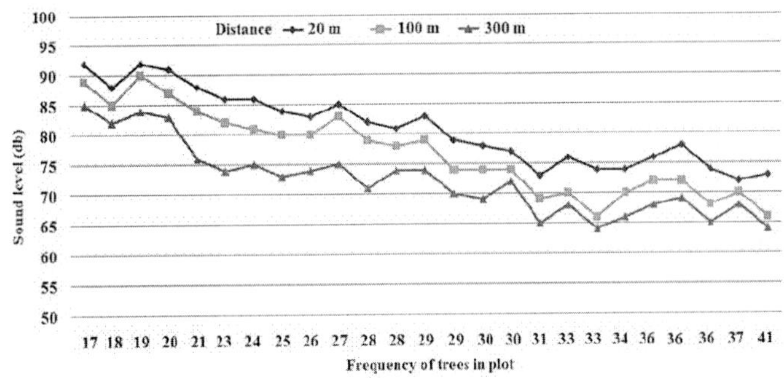

Fig 11 : Comparison of sound level in different distance for broad leaved species

In plots consist of broadleaved species, the sound level at the distances of 20, 100 and 300 meter have been reduced 10.6, 14.5 and 19.4 dB, respectively (Fig. 11).

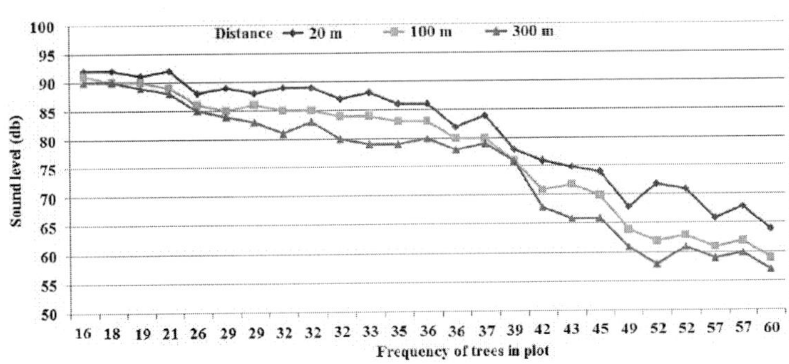

Fig 12 : Comparison of sound level in different distance for coniferous species

This difference with considering these distances was respectively 10.4, 14.3 and 16.8 dB for coniferous species (Fig. 5).

The rate of noise level reduction was not the same in different distance and species. The reason of the different effects of distance on noise pollution reduction in the study area can be explained with the roles of branches and leaves of trees (broadleaved) as barriers to reflect, refract, scatter and absorb sound waves.

Plants leaves absorb acoustic energy by transferring the kinetic energy vibrating air molecule in a sound field to the vibration pattern of the leaves.

Conclusion

Results of this study indicated that the rate of noise polloution increased with increasing frequency of trees and distance from the sound resource. Therefore, the following suggestion can be usefull in order to reduce the sound level at the edge of forest road:

1. Selection of the species must be done based on ecological condition of the site, natural incident (flood, fire), and various pollutions and aesthetic requirements (single type of tree).
2. Plant trees as close together as possible, but leave enough space to allow for healthy growth and development. Some of the coniferous species such as the *Tuja orientalis* which is grown dense (the H:D ratio is low) can stable against noise at a continues wall.
3. It is suggested that the species with low H:D ratio and wide crown used to decrease noise pollution.

Dust Pollution

Dust particles vary in size from visible to invisible. The smaller the particle, the longer it stays in the air and the further it can travel.

Dust particles small enough to be inhaled may cause

- Irritation of the eyes
- Coughing
- Sneezing
- Hayfever
- Asthma attacks

Paper 7: Role of trees in mitigating the problem of noise pollution in stone quarries- A case study of Bengaluru and kolar districts (Naik *et al.,* 2006)

Objective

- To estimate the amount of deposition of particulate on the surface of leaves of various species.
- To study the dust filtering capacity of different species of plants having different leaf shape, physiology etc.
- To suggest suitable plant species for plantation in stone quarry area.

Table 17 : List of plant species and their taxonomical characters selected for the study

Plant Species	Leaf Type	Leaf shape	Family	Leaf nature & Surface	Leaf & Branch
Dalbergia Species	Deciduous, simple	Lanceolate	Myrteceae	Thin, smooth	Small & flexible branches,
Eucalyptus globulus	Evergreen, simple			Thick, smooth	Borne on small drooping branches freely swingling leaves
Acacia auriculifomiis	Evergreen, simple	Ovate	Fabaceae	Thick smooth	Borne on small branchlets, highly drooping freely movable.
Azadiracta indica	Decidious	Ovate	Meliaceae	Small, smooth	Borne on small branches elevated in all direction freely immovable
Grandis	Decidious	Ovate	Verbenaceae	Big broad hairy	Stiff branches, elevated mostly horizontally, freely movable
Ficus religiosa	Evergreen, simple	Chordate	Moraceae	Thin, smooth	Borne on stiff branches with flexible petiole
Bauhinia Varigata	Decidious, simple	Obchordate	Caesalpinaceae	Broad, glabrous	Stiff branches, mostly horizontal
Cassia fistula	Decidious, compound pineapple	Ovate	Caesal pinaceae	Small pinnae, smooth	Stiff branches, leaf axis Flexible, elevated
Millingtonia hortensis	Deciduous compomd	Awl shaped pinnate	Bignonaceae	Smooth, pinnate	Borne on small drooping branch lets, movable.
Pongamia pinnata ensis	Decidious, simple	Ovate	Fabaceae	Brood, smooth	Stiff branches, elevated in all directions Ficus bengal-
	Evergreen, simple	Ovate	Moraceae	Thick, smooth	Borne on stiff branches, elevated mostly horizontal.
Ficus bengalensis	Evergreen, simple	Ovate	Moraceae	Thick, smooth highly margin	Borne on stiff branches, elevated mostly horizontal.
Grevillea robusta					
Tabubia argentea	Semi evergreen palmate	Oblong	Bignonaceae	Thick, smooth	Leaf axis flexible, elevated all direction freely movable.
Pithacolobium dulse	Decidious pinnate	Palmate	Mimosaceae	Small, smooth spinous type	Thin & flexible leaf axis elevated in all directions, movable
Polyalthia longifolia	Evergreen, simple	Lanceolate	Aimonaceae lanceolate	Smooth, highly drooping	Borne on stiff & drooping branch lets, leaf freely immovable.
Terminalia arjuna	Evergreen, simple	Lanceolate	Combretaceae	Smooth, thick	Stiff branches
Mangifera indica	Evergreen simple	lanceolate	Anacardiaceae	Thick, smooth	Stiff branches, elevated in all direction

The detailed morphological characteristics of the sampled plant species in the study area are shown in the Table 18.

Table 18 : Plants arranged according to the weight of the pollutants deposited in descending order

Plant species	Average area of leaves (sq.m)	Avg weight of the pollutant (sq.m)	Standard error
Terminalia arjuna	0.002885	2.657	10.064
Tabubia argentia	0.006645	2.650	10.052
Eucalyptus globulus	0.0062382	2.546	10.046
Mangifera indica	0.0062382	2.468	10.032
Ficus bengalensis	0.0062224	1.842	10.024
Ficus religiosa	0.0061165	1.012	10.020
Acacia auriculiformis	0.0061254	0.946	10.020
Cassia fistula	0.005917	0.843	10.018
Tectona grandis	0.005832	0.769	10.016
Psidium guajava	0.0051708	0.752	10.012
Dalbergia sissoo	0.0057262	0.654	10.010
Grevillea robusta	0.0054262	0.532	10.010
Azadiracta indica	0.0053148	0.419	10.008
Millingtonia sps	0.0052112	0.326	10.008
Polyalthia longifolia	0.0052018	0.228	10.007

The maximum dust depositions recorded were in *Tectona grandis, Ficus bengalensis, Psidium gaujava, Butea monosperma, Mangifera indica, Muntingia calabura, Artocarpus heterophyllus etc*., which may be attributed to hairy surface of leaves, horizontal elevation, stiff petiole and branches etc., which results in less movement of leaves thus exhibiting lesser chances of falling of accumulated dust. Smooth and flexible leaves like *Millingtonia hortensis, Azadiracta* indica, *Melia rubia* do not hold dust to the same degree compared to horizontally arranged leaves of *Grevillea robusta, Delonix regia, Anacardium occidentale, Tamarindus indi7ca species* bearing stiff, horizontal and elevated leaves with rough and hairy surface which hold large amount of dust particles. Similarly drooping leaves and branches have higher chances of dropping down dust during disturbance by agents like wind, animal etc. Free hanging leaves as well as swirling leaves are the main morphological characteristics of dust escaping efficiency as in *Dalbergia sisso.*

Conclusion

The study indicated that, plants can be used as potential device of dust remover in stone quarry area. For dust abatement purpose the leaves showing evergreen, simpled, rough and hairy and fast growing trees are suggested. The above findings suggest that species like *Terminalia arjuna, Tabebuia argentia, Eucalyptus globulus, Muntingia calabura, Ficus bengalensis, mangifera indica, psidium guajava* can be grown in dust polluted areas to reduce the pollution.

On the basis of above papers, the following table indicates plant suitable for pollution abatement

Table 19 : List of plants for pollution control

Air pollution	Noise pollution	Dust pollution
Pithecolobium dulce	Alstonia scholaris	Bougainvillea
Ficus infectoria	Azadiracta indica	Nerium oleander
Mangifera indica	Butea monosperma	Murraya paniculata
Acaia arabica	Grevillea pteridifolia	Cassia fistula
Grevillea robusta	Grevillea robusta	Casurina equisetifolia
Polyalthia longifolia	Tamarindus indicus	Polyalthia longifolia
Psidium guajava	*Terminalia arjuna*	*Tectona grandis*

Carbon Sequestration

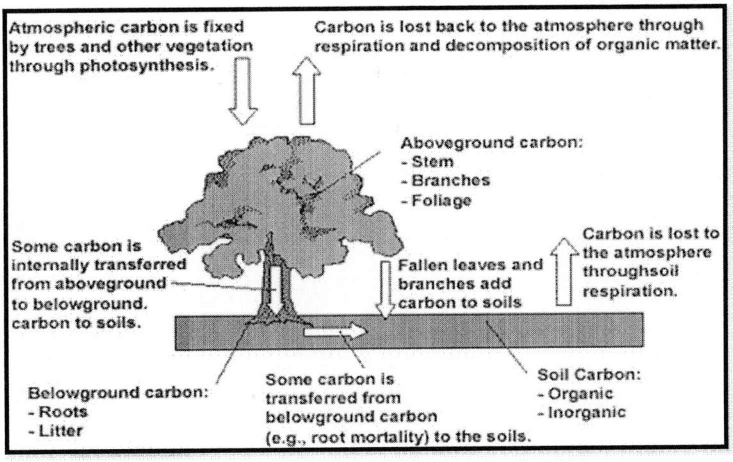

Carbon sequestration is capture and storage of carbon dioxide by plants and storage of carbon in soil.

Through photosynthesis, atmospheric carbon is fixed by trees and other vegetation. Some carbon is internally transferred above groung to below ground. Fallen leaves and branches add carbon to the soil and some carbon is transferred from below ground carbon. Though, some carbon is lost to the atmosphere through soil respiration.

Paper 8: baseline assessment of carbon sequestration potential of existing plantation along national highways of india (Shinge, 2015-17)

Objective: Conduct a baseline assessment of carbon sequestration potential of existing plantation along National Highways of India. Different methods of carbon sequestration.Total carbon sequestered along the highways.

Table 20: Study area details

S. No	National highway	PIU/PD	No. of sampling sites	Soil samples
1.	NH-2	Faridabad/ Mathura	3	3
2.	NH-71A	Rohtak	7	7
3.	NH-31	Guwahati	6	3
4.	NH-37	Guwahati	6	3
5.	NH-40	Shillong	6	3
6.	NH-4	Dharwad	6	8
7.	NH-7	Nagpur	6	3
8.	NH-544	Palakkad	6	3
9.	NH-6	Nagpur	6	4
10.	NH-65	Hyderabad	6	5
		Total	58	42

The sequestration was carried out in 10 national highways using three different assessment method

Table 21: Carbon sequestration calculation using method 1

S.No	National highway	PIU/PD	Total carbon sequestered (Tonnes per acre)
1.	NH-2	Faridabad/ Mathura	No avenue plantation
2.	NH-71A	Rohtak	22.79
3.	NH-31	Guwahati	12.00
4.	NH-37	Guwahati	13.44
5.	NH-40	Shillong	8.32
6.	NH-4	Dharwad	16.64
7.	NH-7	Nagpur	11.16
8.	NH-544	Palakkad	No avenue plantation
9.	NH-6	Nagpur	9.11
10.	NH-65	Hyderabad	9.68

This method was studied using Diameters at Breast height (DBH) and the height of the trees/woody plantation and following:

AGTB= V × D × BEF × CF

BGTB = AGTB x 0.45

SOC= þ*d* % of C

Table 22: Carbon sequestration calculation using method 2

S.No.	National highway	PIU/PD	GBH (m)	No. of trees in sample plot	Basal area	Total C sequestered (tonnes per acre)
1.	NH-2 Mathura	Faridabad/	-	-	-	0
2.	NH-71A	Rohtak	0.80	23	26.96	28.76
3.	NH-31	Guwahati	053	32	22.58	24.06
4.	NH-37	Guwahati	0.63	27	22.74	24.24
5.	NH-40	Shillong	0.67	20	14.22	15.07
6.	NH-4	Dharwad	0.61	15	6.23	6.48
7.	NH-7	Nagpur	0.56	18	8.11	8.51
8.	NH-544	Palakkad	-	-	-	0
9.	NH-6	Nagpur	054	10	2.28	2.24
10.	NH-65	Hyderabad	0.57	21	11.21	11.83

Second method is based on sasal area using the following formulae:
Basal Area (m2) = (GBH/2ð)² *ð
Standing Woody Biomass, SWB (ton/ha) = -1.689+8.32*BA
Below Ground Biomass, BGB = 0.26*SWB
Total biomass= SWB + BGB
Carbon Sequestration= 0.46*Total Biomass

Table 23: Carbon sequestration calculation using method 3

S. No	National highway	PIU/PD	Diameter	Height	Number of trees	Total C sequestered (tonnes per acre)
1.	NH-2	Faridabad/ Mathura	-	-	-	0.00
2.	NH-71A	Rohtak	0.25	9.11	24	28.96
3.	NH-31	Guwahati	0.17	7.875	32	14.44
4.	NH-37	Guwahati	0.20	9.60	30	23.36
5.	NH-40	Shillong	0.21	8.10	20	14.97
6.	NH-4	Dharwad	0.19	9.78	15	11.30
7.	NH-7	Nagpur	0.18	6.66	18	7.80
8.	NH-544	Palakkad	-	-	-	0.00
9.	NH-6	Nagpur	0.17	6.57	10	3.90
10.	NH-65	Hyderabad	0.18	8.03	21	11.14

This method is based on diameter and height of the tree using the formulae:
Bio-volume (T) =0.4X (D) 2 * H
Above Ground Biomass =Wood density* T
BGB=AGB x 0.26
Carbon Storage = Biomass x50%

Table 24: Carbon sequestered by median plantation and soil organic carbon

S.No	National highway	PIU/PD	Total C sequestered (tonnes per acre) Median	Soil organic carbon (%)	Tree cover/ plantation
1.	NH-2 Mathura	Faridabad/	0.115	2.56	57.38
2.	NH-71A	Rohtak	0.272	1.83	60.00
3.	NH-31	Guwahati	0.39	2.56	47.00
4.	NH-37	Guwahati	.523	5.63	46.66
5.	NH-40	Shillong	.070	1.97	69.00
6.	NH-4	Dharwad	0.230	4.69	65.00
7.	NH-7	Nagpur	0.30	3.14	41.00
8.	NH-544	Palakkad	0.207	3.89	63.00
9.	NH-6	Nagpur	0.292	4.03	95.00
10.	NH-65	Hyderabad	0.260	2.87	64.00

The total carbon sequestered including median and soil organic carbon was calculated and tree cover percentage was found.

Although all methods have their applicability in different scenarios, method 1 was found to be most significantly preferable (Table 24).

Table 25: Parameters of three methods used

Parameter	GBH/ Diameter	Height	Number of trees	Species specific	Climatic condition
Method 1	Y	Y	Y	Y	Y
Method 2	Y	N	Y	N	Y
Method 3	Y	Y	Y	N	N

Table 26: Summary of mean carbon sequestration estimated in the 3 C pools

National highway	Agro Climatic zone	Total C sequestered (tonnes per acre)			Total sequestration (tonnes per acre)
		Avenue	Median	Soil organic carbon	
NH-2, NH-71A	Trans gangetic plain	11.39	0.19	2.22	13.80
NH-31, NH-37 NH-37, NH-40	Eastern himalayan region	11.25	0.33	3.39	11.58
NH-65, NH-4 NH-7 NH-544	Southern plateau and hills region	9.37	0.25	3.65	13.27
NH-4, NH-7 NH-544, NH-6	West coast plain and ghat region	9.23	0.26	3.94	13.42

Trans Gangetic plain showed the highest carbon sequestration as compared to other regions. Hence, average area under plantation and total carbon sequestered was calculated and summarized in Table 27.

Table 27 : Summary of zone wise total carbon sequestered

Agro Climatic zone	States/ region	Vegetation cover (%)	Total stretch (km)	Total C sequestered (tonnes per acre)			Total sequestration (tonnes per acre)
				Avenue	Median	Soil organic carbon	
Trans gangetic plain	Delhi, Chandigarh, Punjab, Haryana	58.69	5486.63	90654.83	615.02	42090.12	133359.98
Eastern himalayan region	North Eastern region	54.22	12691.91	191362.24	2244.96	64288.18	257895.38
Southern plateau and hills region	Andhra Pradesh, Karnataka, Tamilnadu	58.25	7118.55	96024.77	1026.12	69210.62	166261.52
West coast plain and ghat region	Maharashtra, Madhya Pradesh	66.00	6314.18	95033.96	1064.92	74715.06	170813.95
Total			31611.27			728330.82	

Vegetation cover is first in western plateau and hills region i.e., 66% but the carbon sequestration is low compared to other regions , because the plantation done here are fresh and premature.

Conclusion

Total area under national highways is 1.00.089 km which can sequester 2.13 to 2.46 MT. This shows the effectiveness of national highways and has a great potential to be an effective and economic way to sequester carbon.

Paper 9: Quantitative analysis of carbon storage of 25 valuable tree species of gujurat, Incredible India (Pandya *et al.,* 2013).

Objective: To quantify the carbon storage in few selective species or heritage tree species of Gujarat.

Table 28: Selected tree species and district wise their physiological details

Vernacular name	Scientific name	GBH (m)	Height (m)
Largest rukhado tree	*Adinsonia digitata*	11.3	20
Largest kadaya tree	*Sterculia urens*	3.86	25
Largest mango tree	*Mangifera indica*	7.4	16
Largest semal tree	*Bombex ceiba*	13.5	35
Largest Amla Tree – Mahavriksh	*Emblica officinalis*	3.4	10
Largest Neem Tree	*Azadiracta indica*	6.9	24
Largest Baheda Tree	*Terminalia bellirica*	6.15	39
Tallest Sadad Tree	*Terminalia tomentosa*	3.05	36
Largest ArjunSadad Tree	*Terminalia arjuna*	7.30	35
Large Haldu Trees	*Adina cordifolia*	4.20	35
Old and Largest Bordi Tree	*Ziziphus mauritiana*	4.68	15
Old Amli Tree	*Tamarindus indica*	7.55	30
Large and Old Rayan Tree	*Manilkara hexandra*	6.50	30
Tallest Teak Tree	*Tectona grandis*	4.21	41.5
Largest Rain Tree	*Samanea saman*	5.56	41
Largest Kalam Tree	*Mitragyna parviflora*	4.10	41.5
Tallest Kilai Tree	*Albizia procera*	2.30	38
Largest Mahuda	*Madhuca indica*	6.05	32
Tallest Umbh Tree	*Miliusa tomentosa*	2.30	37
Oldest Piloo Tree	*Salvadora oleoides*	6.45	12
Largest Mahogany Tree	*Swietenia mahogani*	6.80	30
Oldest Pipal or Pipado Tree	*Ficus religiosa*	9.40	16
Large Pipada Tree	*Ficus tsila*	6.40	24.7
Oldest and Largest Borsalli Tree	*Manilkara elengi*	4.10	18
Largest Shivlingi Tree	*Couroupita guianensis*	3.55	23

Table 29 indicates the list of species selected for the study. The girth at breast height and diameter of the tree were measured.

Table 29: List of species and their above, below, ground and total biomass with carbon

Scientific name	AGB (kg)	BGB (Kg)	TB (Kg)	C (Kg)	tC/species
Tamarindus indica	88802.63	23088.68	111891.31	55945.66	55.95
Terminalia arjuna	71128.28	18493.35	89621.63	44810.82	44.81
Bombex ceiba	64695.83	16820.92	81516.75	40758.37	40.76
Manilkara hexandra	54507.28	14141.89	68679.18	34339.59	34.34
Terminalia bellirica	45480.91	1825.04	57305.94	28652.97	28.65
Madhuca indica	35163.58	9142.53	44306.12	28652.97	22.15
Samanea saman	30852.15	8021.56	38873.70	22153.06	19.44
Azadiracta indica	30595.25	7954.77	38550.02	19436.85	19.28
Swetinia mahogany	28701.85	7462.48	36164.34	19275.01	18.08
Adinsonia digitat	25408.26	6606.23	36030.24	18082.17	18.02
Ficus religiosa	24262.31	6402.98	32014.79	18015.12	16.01
Ficus tsila	24214.21	6284.44	31029.83	16007.39	15.51
Mangifera indica	24170.91	5586.23	30455.35	15549.91	15.23
Tectona grandis	21485.50	4709.45	27071.73	15227.67	13.54

Contd.

Mitragyna parviflora	18113.25	3842.31	22822.70	13535.87	11.41
Adina cordifolia	14778.12	3159.56	18620.44	11411.35	9.31
Salvadora oleodies	12152.14	2133.48	15311.69	9310.22	7.66
Sterculia urens	8205.69	2119.47	10339.17	7655.85	5.17
Terminalia tometosa	8151.81	2079.26	10271.28	5169.59	5.14
Ziziphus mauritiana	7997.14	1914.99	10076.40	5038.20	5.04
Manilkara elengi	7365.33	1437.00	9280.31	4640.16	4.64
Couroupita guianensis	5526.92	1272.23	663.92	3481.96	3.48
Albizia procera	4893.18	1238.75	6165.41	3082.70	3.08
Miliusa tomentosa	4764.41	1211.58	6003.16	3001.58	3.00
Emblica officinalis	2813.91	731.62	3545.52	1772.23	1.77

This table shows in descending order the carbon stored in the plants. It is observed that *Tamarindus* indicus showed the highest storage of carbon whereas *Emblica officinalis* had the lowest carbon stored in it. It was observed that the plants with highest girth and diameter had the highest carbon.

Conclusion

This indicates that the plants as they age have the highest storage capacity. So, the plants have to be grown and maintained to capture the highest carbon in them.

Urban Heat Islands

Heat Islands develop when a large fraction of the natural land cover in an area is replaced by built surfaces that trap incoming solar radiation during the day and then re-radiate it at night (Quattrochi *et al.,* 2000).

Heat islands has various secondary effects on local meteorology including wind, Rh and precipitation. It results in climate warming by 30% and it also results in increasing electricity demand.

Paper 10: Tree canopy cover variation effects on urban heat island in Enugu city, Nigeria (Enete *et al.,* 2012).

Objective

i. Quantify actual temperature benefits of street trees

ii. Assess the benefit of different tree canopy

iii. Quantify thermal comfort benefits of some selected street trees.

Table 30: Mean monthly temperature management

Month	Non-tree temperature	Under tree temperature
February	31.0°C	23.0°C
March	32.5°C	27.5°C
June	29.8°C	17.8°C
July	28.2°C	18.2°C
Feb- July	30.3°C	21.6°C

The mean monthly temperature for dry season represented by February and march were 31 and 32.5 degree Celsius while june july 29.8 and 28.2 temperature was recorded and under tree 23.0, 27.5 and 17.8, 18.2 respectively (Table 29). On an average February had 30.3°C while total temperature under foliage was 21.6°C (Fig 14).

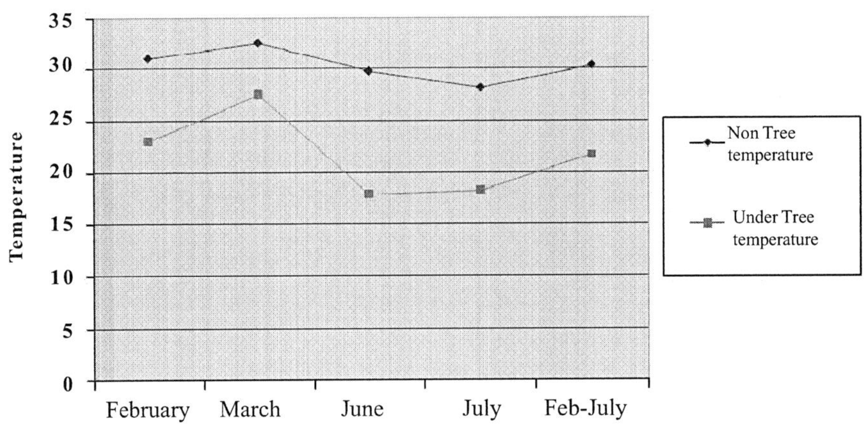

Fig. 14 : Mean monthly temperature management

Table 31: Mean temperature under different trees

Tree/ plant	Mean temperature
Anacardium (Cashew)	12°C
Citrus genus	18°C
Cocus nucifera	28°C
Mangifera indica	14°C
Psidium guajava	17°C
Azadiracta indica	19°C
Pinus genus	25°C
Catalpa bungei	12°C

Highest temperature under tree foliage was recorded in coconut followed by Pinus, whereas Catalpa and Anacardium recorded the lowest mean temperature. The thermal comfort benefit was determined using discomfort index by Tzenkova *et al.* (2000).

Table 32: Classification of discomfort index (Tzenkova *et al.*, 2000)

Thermo-hydrometric discomfort index	t°c
No discomfort	<21
Under 50% of population feels discomforted	21-24
Over 50% of the population feels discomforted	24-27
Most of the population feels discomforted	27-29
Everyone feels stressed	29-32
State of medical emergency existed	>32

The analysis of temperature of selected tree revealed that 75% of the trees did not cause any discomfort, whereas coconut and Pine came under where over 50% of the population feels discomforted.

Conclusion

A fully solar gain of < 20 BTU/m/ft^2 therefore shading of tree eliminates 90% of solar energy falling on surface. This is because trees have a cooling effect due to evapotranspiration.

Here are the lists of the species according to the space available for plantaion on avenues.

Table 33: Species recommeneded for Ist row of avenue plantation

Soil	Botanical name	Local name
Loamy	*Delonix regia*	Gulmohar
	Cassia fistula	Amaltas
	Bahunia sps.	Kachnar
	Cassia nodosa	Cassia
	Jacaranda mimosifolia	Jacaranda
	Peltophorum ferrugineum	Copper pod
Water logged condition	*Terminalia arjuna*	False almond
	Syzygium cuminii	Jamun
	Cordia dichotoma	Lasoda
Alkaline soils	*Terminalia arjuna*	False almond
	Pongamia pinnata	Karanj
	Albizzia lebbeck	Kala siris

Table 34: Species recommended for 2nd and subsequent row, except the last row of avenue plantations

Soil	Botanical name	Local name
Loamy	*Melia azadiracta*	Neem
	Pongamia pinnata	Karanj
	Grevillea robusta	Silver oak
	Albizzia lebbeck	Kala siris
	Dalbergia sissoo	Shisham
	Teminalia arjuna	False almond

Table 35: Species recommended for last row of avenue plantations

Soil	Botanical name	Local name
Loamy	*Ficus religiosa*	Peepal
	Ficus infectoria	Paker
	Madhuca indica	Mahua
	Mangifera indica	Mango
	Azadiracta indica	Neem
	Tamarindus indica	Imli
Water logged condition	*Syzygium cuminii*	Jamun
	Terminalia catapa	Arjun
Alkaline soils	*Azadiract indica*	Neem
	Pongamia pinnata	Kanji
	Terminallia catapa	Arjun
Sandy	*dalbergia sissoo*	Shisham

Future

It can be used to reduce the impact to natural habitats

It is essential in initiating mitigation initiatives to reduce carbon emission

It also requires adaptation initiatives to reduce economic and environmental damages

Conclusion

Green corridors would strike balance between development and environmental protection, and will go long way in creating greener corridors for comfortable road journey, keeping environmental amelioration in place. It would encourage buy back arrangements for sustainable harvest, convergence of other government schemes, enhancing people's participation in greening highways and carbon emission trading. It will provide self employment oppurtunities like creation of nursery, tree guard manufacturing and agri-processing sectors. NGHM is expected to develop 4 lakh acre land which will help in sesquetering 12 MT carbon annually. This comes at a time when concern for neutralising ill effects of pollution is shared globally and would help India in 35% reduction in carbon emissions by 2030.

With the development of green corridor, travelers would be able to enjoy the beauty and extensive benefits of the vegetation, while experiencing safer driving environments. Green corridor development will also help in achieving the goal of maintaining one third of geographical area of the country under tree cover.

References

Bachman, P.K. (2016). Report. Retrieved from stateoftheworldsplants.com: https://stateoftheworldsplants. com/report/sotwp_2016.pdf

CPCB, (2010) Central Pollution Control Board, New Delhi., 50-51.

Daucus, T., (2011). Landscape Maintenance Zone, Distances and Heights. *J. Env. Bio.,* 30(4): 545-550.

Enete, I. C., Alabi, M. O. and Chukwudelunzu, V. U. (2012). Tree canopy cover variation effects on urban heat island in Enugu city, Nigeria. *Intl. Inst. Sci. Tecnol. Edu.,* 2(6):12-18.

Floriculture Today., (2016). p. 26.

Gupta, S., Nayek, S. and Bhattacharya, P. (2012). Effect of air-borne heavy metals on the biochemical signature of tree species in an industrial region, with an emphasis on anticipated performance index. *Chem. Eco.,* 27(4): 381-392.

Intergovernmental Panel for Climate Change, 2015

Mate, A. R. and Deshmukh, R.D., (2015). To control the effect of air pollution using roadside trees. *Intl. J. Inn. Res. Sci. Engg. Technol.,* 4(11):11167-11172.

Mok, J., Landphair, H. C. and Naderi, J. D., (2006). Landscape improvement impacts on roadside safety in Texas. *Lscp. Urban Plng.,* 78: 263-274.

Naik, P., Ushamalini. and Somashekhar, R. K., (2006). Role of trees in mitigating the problem of noise pollution in stone quarries- A case study of Bengaluru and kolar districts. *J. Indstrl. Poll. Ctrl.,* 22(2): 291-296.

Nasiri, M., Fallah, A. and Nasiri, B., (2015). The effects of tree species on reduction in rate of noise pollution at the edge of hycanian forest roads. *Env. Engg. Mgmt. J.,* 14(5):1021-1026.

Nowak, D. J., (2002). The effects of urban trees on air quality. *USDA Forest Service.,*1-5.

Onder, S. and Kockbecker, Z. (2012) Importance of green belts to reduce noise pollution and determination of roadside noise reduction effectiveness of bushes in Konya, Turkey. *Intl. J. Biol. Biomol. Agri. FoodBiotechnol.,* 6(6): 373-376.

Pandey, J., Pandey, R. and Shubhashish, K., (2009). Air-borne heavy metal contamination to dietary vegetables: a case study from India. *Bull. Environ. Contam. Toxicol.,* 83: 931–936.

Pandya, I. Y., Salvi, H., Chahar, O. and vaghela, N., (2013). Quantitative analysis of carbon storage of 25 valuable tree species of gujurat, Incredible India. *Indian J. Sci. Res.,*4(1): 137-141.

Quattrochi, D., Luvall, J., Rickman, D., Estes, M., Layman, C. and Howell, B. (2000). A decision support information system for urban landscape management using thermal infrared data. *Photogram. Engg. Remote Sensing.,* 66(10):1195-1207.

Rai, P.K. and Panda, L.L.S. (2015). Roadside plants as bio indicators of air pollution in an industrial region, Rourkela, India. *Intl. J. Adv. Res. Technol.,* 4(1): 14-28.

Royal Botanic Gardens, Kew, UK., 2016, Retrieved from stateoftheworldsplants.com: https://stateoftheworldsplants.com/

Sandermann, H., Ernst, D., Heller, W. and Langebartels, C. (1998). Ozone: an abiotic elicitor of plant defence reactions. *Trends. Plant. Sci.,* 3: 47-50.

Sharma, R.K., Agarwal, M. and Marshall, F.M., (2008). Heavy metal (Cu, Zn, Cd and Pb) contamination of vegetables in urban India: a case study in Varanasi. *Environ. Pollut.* 154: 254–263.

Sharma, S.J., (2011). Growth rate of Motor Vehicles in India - Impact of Demographic and Economic Development. *J. Eco. Soc. Studies.,* 3(2): 137-150.

Shinge, A., (2016). Baseline assessment of carbon sequestration potential of existing plantation along national highways of India. PGDFM thesis, Indian Institute of Forest Management, Bhopal.

Tripathi, A.K. and Gautam, M., (2007). Biochemical parameters of plants as indicators of air pollution. *J. Env. Biol.,* 28(1):127-132.

Tripathi, A., Tiwari, P. B., Mahima. and Singh, D., (2011). Assessment of APTI of some trees in Moradabad, India. *J. Env. Biol.,* 30(4):545-550.

Tzenkova, A., Kandjou, I. and Ivancheva, J. (2000). Some Biometeorological Aspects of Urban Climate in Sofia. *Scientists contributions Journal, Eurasap.*

Vision document- strategy, mission, action plan. NGHM, MoRTH, 2015.

www.cpcbenvis.nic.in

www.forestry.gov.uk

7

Roof Gardening

Sachin Tyagi[1], K. Rama Krishna[2], Jyoti Kumari[1]
Preeti Kumari[3] and Ankit Singh[4]

[1]Department of Horticulture, Bihar Agricultural University Sabour, Bhagalpur- 813210 (Bihar) India
[2]Division of Food Science and Postharvest Technology, IARI, New Delhi-110012
[3]Department of Horticulture (Vegetable & Floriculture), Bihar Agricultural University, Sabour, Bhagalpur-813210 (Bihar) India
[4]Gandhi Smarak Devnagari Mahavidyalaya Parikshat Garh, Meerut CCS University Meerut-250406 (U.P.) India

Introduction

Any kind of garden established on the roof top of building is known as roof garden, which provides the opportunity to cultivate the quality vegetables, fruits and flowers in home. In this system plants and grasses are grown on a nutrition medium placed above a waterproof membrane. Roof gardening is mostly adopted in urban areas. It uses for decorative benefit, roof plantings make available food, manage temperature, hydrological remuneration, architectural improvement, habitats for wildlife, frivolous opportunities, and in large scale it may even have ecological benefits (Louise, 2009). This considerably increases human health as adding organic cultivation.

Green terrace tops with plants and flowers grant green and cool spaces, energy conservation, best quality air for breath, healthy life, good biodiversity (Jose, 2014). Terrace gardens change the visual look of the building, viewing from neighbors and hide redundant pipeline. Green roof is an important policy that addresses some key urban environmental issues. Therefore, popularity of roof garden is rising in the smart cities across the world.Green roofs are beneficial to reduce surface water overflow as well as providing the habitats for wildlife. Roof gardening plays a good role in to improve air quality and create the visual and amenity value of urban areas. These types of gardens provide the opportunity for food production in urban areas.

Benefits of Rooftop Gardens

- Mitigate the negative impacts of cities on the environment.
- Increased access to safe outdoor green space
- Water and energy conservation.
- Absorption solar radiation.
- Provide the habitat restoration.
- Sound absorption.
- Water absorption.
- Storm water retention
- Provide architectural enhancement.
- Reduction of dust and smog level.
- Reduced noise level.
- Reduce wind speed
- Increase the heat resistance of the roof
- Increase the beauty of cities.
- Provide more space for agriculture.

Types of Roof Garden or Green Roof

Generally, roof gardens are classified in the following three categories.

1. **Extensive:** This type of roof gardens required minimum maintenance among types. The costs are lower than Simple Intensive or Intensive Green Roofs.
2. **Semi Intensive:** Semi-Intensive roof garden requirements drop in the middle of Extensive and Intensive Green Roof systems. More maintenance, higher costs and more weight are the main characteristics of semi-Intensive roof garden compared to extensive roof garden.
3. **Intensive:** Intensive gardens formed with the objective of lively human utilize. Thissystem includes the lawn, perennials, bushes and trees. Some additional features can be established in the intensive roof gardening like walkways, benches, playgrounds or even ponds etc.

Construction of Roof Garden

A green roof is created by adding of different layers on top of a traditional roofing system. The waterproofing of the roof can be remain protected from any mechanical damage and also from direct influence of the external factors. The layers of a contemporary green roof system, from the top down, includes following layers (Fig-1).

1. Vegetation or Plants
2. Growing medium
3. Filter membrane
4. Drainage layer
5. Thermal Insulation
6. Waterproofing/roofing membrane
7. Vapour control layer or Protection board
8. Roof deck or Structural Support

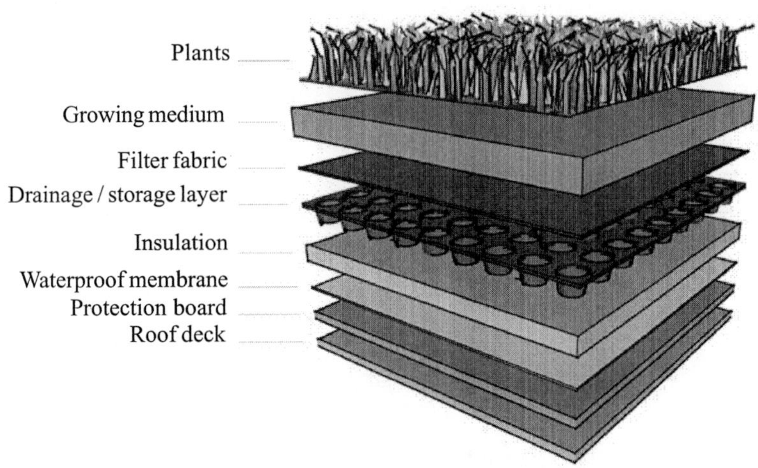

Fig. 1 : Different layers of green roof.
Source: https://myrooff.com/roof-garden-construction

Vegetalization systems of roof gardening

Dobrescu E. (2012) studied on twovegetalizationsystems (Extensive vegetalization and Intensive system vegetalization) of roof gardening to extend the area of green space in urban areas.

1. Extensive vegetalization

This system can be applied to roofs of old buildings expertised to support maximum 250 kg./sqm (Dobrescu E., 2012). Extensive vegetalization includes Easy-load on structural strength, landscaping on any roof surface, small and very small thickness of substrate vegetation (3-10 cm.), very low or no-maintenance and ecological function for improving the microclimate (Dobrescu E., 2012) (Fig.1).

Fig. 1 : Extensive vegetalization terrace

2.Intensive system vegetalization

This system vegetalization can be implemented only on new buildings provided by the project to be vegetalizated, or old ones that after technical expertise are deemed to support loads higher than 300kg./m^2 (Dobrescu E., 2012). The intensive vegetalization includes simply and instant accessibility for people and visitors, large-load resistance structure, multifarious maintenance of irrigation, fertilization, maintenance, well landscaped and gardens or parks on the ground etc (Dobrescu E., 2012) (Fig. 2).

Fig. 2 : Intensive vegetalization terrace

Management of Roof Garden

1. Growing media

Low dry bulk density, high water holding capacity and high air filled porosity are the desired physical characteristics of growing media for green roofs.

2. Soil

Lightweight soils are considered best for roof garden. Perlite, peat moss, vermiculite and coconut husk fibers are mixed with soil to improve water holding capacity.

3. Reduce the load on top roof

The load on top roof can be reduce by using mix soil with organic and inorganic matter or other lightweight growing mediums, such as peat moss, vermiculite or perlite etc.

4. Plant selection

The plants should be select according to climatic conditions and geographical location.The selected plants should be resistant against the biotic and abiotic stress. In the water shortage areas drought tolerant herbs and shrubs performed better. Pinus, Hedera, Junipers and Cotoneaster are good drought tolerant plants for roof gardening. Moisture tolerate plants can perform better in the high rainfall areas. Thus, various types of ornamental plants, shrubs, vegetables, herbs and small fruits can be grown successfully on the green roofs. Some important recommended group of plants for roof gardening are shown in Table (1.).

Table 1 : Recommended Group of Plants for Roof Gardening.

S.No.	Group	Species or plants	Characteristics
1.	Trees	*Caesalpiniacoriaria, Callistemon polandii, C. lanceolatus, Cordia sebestina, C. boissieri, Cycas* spp., *Ficusbenjamina, Jufniperus chinensis, Nyctanthes arbor tristis, Polyalthia longifolia, Plumeria* spp., *Thevetia peruviana, Thuja orientalis, Plumeria alba, Pomegranate, Citrus trees, Palms*	Dwarf tree with colourful flowers.
2.	Shrubs	*Angelonia grandiflora, Acalyphahispida, A. wilkesiana, Beloperone* spp., *Aralia* spp., *Asystasiacoromandeliana, Barleriacristata, Caesalpiniapulcherrima, Calliandrabrevipes, Cestrum* spp., *Crossandrainfundibuliformis, Cuphea miniata, Daedalacenthus azureus, Hibiscus, Jasminum* spp., *Lagerstroemia indica, L. depressa, L.nivea., L. sellowiana, Malpighacoccigera, Pentuscarnea, Plumbagocapensis, Poimsettia, Russeliajuncia.*	Beautiful colour flours or foliage, easy to grow.

3.	Creepers	*Adenocylymmaallicea, Allamandapurpurea, A. cathartica, Bignonia purpurea, B. tweediana, Bougainvillea cvs., Clerodendrumsplendens, Jacqemontiaviolacea, Passiflora spp., Tecomaradicans, Thunbergiamysorensis, Wisteria chinensis.*	Free flowering specimen creepers.
4.	Bulbous	*Canna spp.*, Caladium, Gladiolus, Hrenerocallis, Hippeastrum, Haemanthus, Zephyranthus.	Shallow rooted, colourful flowers.
	Large succulents	*Aloe, Jade plant*	Ability to store water
	Low succulents	*Sedums, Aptiniacordifolia, Portuluca, Crassula* spp	Ability to store water
5.	Annuals	*Alyssum, Antirrhinum, Cosmos*, Marigold, Nasturtium, Petunia, Pansy, Phlox, Poppy, Sweet William, Portulaca, Gomphrena.	Ideal for bedding or potting.
	Grasses	Zoysia –korean grass, Bermuda grass, Ornamental grasses	Contribute to Aesthetic value
	Herbs and vegetables	Thyme, Rosemary, All greens	Provide food.
	Ground covers	Setcreaseapurpurea, Wedelia , gourds protection	purifies the air, Erosion

Source : Roy P.K. (2013) and TNAU, Agritech Portal.

5. Stability

Tap-rooted plants need deep soil. Therefore, dwarf plants with shallow root pattern will be best for roof gardens like Scrub Pine and Birch trees. Thus, these plant can be maintain easily in the roof gardening.

6. Irrigation

In the roof garden water should be applied regularly, as needed. Drip irrigation, Sprinkler irrigation,overhead irrigation system and automated irrigation systems are preferred in roof garden for water application.

7. Drainage

To protect the vegetation from water logging condition, a good water drainage system should be properly maintain on the green roofs. This system can be maintain by using a gravel ring and filter cloth layer around roof drains and overflow scuppers.

8. Weed Control

Weeds must be controlled until the plantings achieve the desired growth rates and/or coverage.

9. Mulching

Drying and overheating of the soil can be prevented by the application of 2-3 inches of shredded hardwood mulch (Anderson M.).

10. Greenhouses

Greenhouses and cold frames protect the plants and grasses from wind, drought and sunlight. An automatic green house also controlled the environmental condition, as needed. Thus, this system on green roof can extend the growing season and provide the year-round production.

Impacts of Roof Gardening

1. Increase life of roof top

The green roofs are constantly improving the roof top isolation (Dumitras *et al.*, 2008). The research show that using a green roof top it was reduce with 26% the need of conditioned air in summer and a 26 % need of heat during winter (Dumitras *et al.,* 2008). Thus, the life time warranty of this type roof tops increases with 2-3 times more (Dumitras *et al.,* 2008).

2. Improvement in microclimate of urban areas

Microclimate of urban can be improved by growing plants with photosynthetic potential.

3. Air Quality

Plants are natural filters of toxins, they are able to trap particles on their leaves, branches, and stem surfaces until it rains and they are washed into the soil (Dam *et al.*, 2000).As air moves across a garden air borne particles and gaseous pollutants are filtered out of the air by the plants (Dam *et al.*, 2000). For every one meter square of grass roof, about 0.2 kg of air borne particles can be removed from the air every year (Velazquez, 2000). Nitrous oxides and volatile organic air pollutants that can be absorbed by the foliage are either sequestered in the leaves or used by the plant for photosynthesis (energy production) (Dam *et al.*, 2000).

4. Reduce energy demand

Roof gardens can reduce energy demand on space conditioning, and hence GHG emissions, through direct shading of the roof, evapotranspiration and improved insulation values (Liu, K., 2014).Wasima and Nine (2016) studied on DOE-2 energy simulation program to determine the effects of rooftop garden on the annual energy consumption of a ten story residential building in Dhaka, Bangladesh. This research has shown that the installation of rooftop garden in a ten story residential building at Bangladesh could result in a saving of 2 to 8% of total annual energy consumption (Wasima and Nine, 2016). Thus, a green roof can reduce energy consumption.

5. Improved thermal performance

Green roof, also reduce the heating and cooling requirements of a building. Green roofs reduce heat transfer through the roof andambient temperatures on the roof surface, improving the performance of heating, ventilation and air conditioning (HVAC) systems (*Growing Green Guide*, 2014).

6. Reduce the urban heat

Green roofs are effective way to decrease urban temperatures due to their versatile effects and functions, such as isolation or runoff peak delay. If widely adopted, rooftop gardens could reduce the urban heat island, which would decrease smog episodes, problems associated with heat stress and further lower energy consumption (Liu, K., 2014).

7. Carbon Sequestration

Ornamental landscapes have good potential for carbon sequestration. Whittinghill *et al.* (2014) reported that ornamental landscapes both in-ground and on green roofs have the ability to sequester carbon.

8. Food production in urban areas

Different types of fruit, vegetable and grain can be grown on green roof. Thus, it can play a good role in food production in urban areas.

9. Storm water management

Green roofs are a popular storm water management technique because they are effective in reducing runoff volume from buildings (Malcolm *et. al.*, 2014).Green roofs absorb and retain rainwater and can be used to manage storm water run-off in urban environments (Growing Green Guide, 2014).They can also filter particulates and pollutants. Storm water run-off can be reduced or slowed because it is stored in the substrate, used by or stored in the foliage, stems and roots of plants, and also evaporates directly from the substrate(Growing Green Guide, 2014).

10. Creation of habitat & ecological biodiversity

Green roofs increase urban biodiversity by providing habitat for wildlife. They also provide a link or corridor across urban 'ecological deserts' and assist in migration of invertebrates and birds (Growing Green Guide, 2014).

Conclusions

Green roof has substantial potential to increase the sustainability of urban areas. Green roof can reduce energy consumption, noise level, and urban heat. Ornamental landscapes have good potential for carbon sequestration. Plants can filter the toxic pollutant from the air. Green roof can increase the life time of roof top. It also enhance biodiversity by providing new urban habitats. Thus, Roof gardening improve the environment and image of urban areas, increase urban biodiversity and provide the new opportunities for quality food production.

References

Anderson M.. Design and development of a roof garden. *Rooftop Garden Manual*. www.mccaren.com.

Berar C., Băla Maria, Fora C.G. and Cristina Toa. (2012). Researches concerning landscape development project of a roof garden in Timisoara municipality.*Journal of Horticulture, Forestry and Biotechnology.* 16 (4):77- 84.

Brahic C. (2007). Cooling percentages by 'Green roofs'". *New Scientist*. Retrieved March 12, 2014.

Dam A., Walke C. and Wilson C. (2000). The Future of Rooftop Gardens on the University Of Waterloo Campus. http://www.grow.uwaterloo.ca/Gardens.pdf.

Dobrescu E. (2012). Green roofs/roof gardens. Research on roof/terrace vegetalization; vegetalization systems. Scientific Papers, Series B, Horticulture. Vol. LVI. ISSN.2285-5661.

Dunnett N. and Kingsbury, N. (2008). Planting green roofs and living walls, 2nd edition. Timber Press, Portland Oregon.

English Nature. (2003). Green Roofs: Their existing status and potential for conserving biodiversity in urban areas. English Nature Report No. 498. English Nature, Peterborough.

FAO (Food and Agriculture Organisation). (1992). Waste water treatment and use in agriculture. FAO Irrigation and Drainage Paper 47, FAO, Rome. Retrieved 2010 from http://www.fao.org/docrep/t0551e/t0551e 00.HTM.

JoseV. P. (2014). Balcony and terrace gardens in Urban greening and local food production: Scenarios from Mumbai Metropolitan Region (MMR), India. *Int. J. FoodAgri.Vet. Sci.* 4:149-162.

Kumar R. and Kaushik S.C. (2005). Performance evaluation of green roof and shading for thermal protection of buildings, *Energy and Buildings*, 27:1505-1511.

Liu, K. (2014). Energy efficiency and environmental benefits of rooftop gardens. *National Research Council Canada*. Retrieved March 12,2014.

Louise Lundberg Scandinavian Green Roof Institute (2009). "The benefits of Rooftop Gardens" (PDF). Retrieved March 12, 2014.

Malcolm E. G., Reese M. L., Schaus M. H. ,Ozmon I. M. and Tran L.M. (2014). Measurements of nutrients and mercury in green roof and gravel roof runoff. *Ecological Engineering* 73 (2014) 705–712.

Mumitras A., Ilca T.S., Singureanu V., Sabo G.M., Pop P. and Moldovan G. (2008). Green roofs in urban landscape. Bulletin UASVM, *Horticulture* 65(1):487-487.

Navid V. , Ali T.andFatemeh K. (2016). Energy Conservation Potential of an Extensive Green Roof in Iran for One Year Duration. *Journal of Ornamental Plants*. ISSN 2251-6433.

Ould-Ahmed B.A., Yamamoto T., Rasiah V., Inoue M. and Anyoji, H. (2007). The impact of saline water irrigation management options in a dune sand on available soil water and its salinity. *Agricultural Water Management*, 88:63-72.

Rhoades J.D. (1998). Use of saline and brackish water for irrigation, implications and role in increasing food production, consuming water, sustaining irrigation and controlling soil and water degradation. Proc of the International Workshop on the Use of Saline and Brackish Water for Irrigation, At 10th Afro-Asian Confr. Bali, Indonesia, 261-304.

Roy P.K. (2013).Roof Gardening. Fundamentals of Garden Designing. New India Publishing Agency, 617-620.

V. Pandiyaraju, P. ShunmugaPerumal, A. Kannan and L. Sai Ramesh. (2017). Smart terrace gardening with intelligent roof control algorithm for water conservation. *Pak. J. Agri. Sci.,* 54(2):451-455.

Velazquez, Linda S. (2000). "Greenroof Components." Greenroofs.com: Exploring the ecology of organic greenroof architecture. http://www.greenroofs.com/components.htm.

Wasima J. and Nine A.H. J. (2016). Adapting cities for climate change: an assessment of green roof technology. *International Conference on Anthropology, Adaptation and Resilience in Climate Change Regime.*, 22-23 October 2016 University of Dhaka, Dhaka-1000, Bangladesh.

Whittinghilla L.J., Roweb D. B., Schutzkic R., Creggd B. M. (2014).Quantifying carbon sequestration of various green roof and ornamental landscape systems. *Landscape and Urban Planning*. 123. 41-48.

Growing Green Guide (2014). A guide to green roofs, walls and facades in Melbourne and Victoria, Australia.www.growinggreenguide.org .

http://www.igra-world.com/types_of_green_roofs/16/10/2017.

https://myrooff.com/roof-garden-construction

http://agritech.tnau.ac.in/horticulture/horti_Landscaping_roofgarden.html /20/10/2017.

https://www.evergreen.ca/downloads/pdfs/Design-Ideas-4-Rooftop-Gardens.pdf

8

Beekeeping as a Fifth and Most Important Input for Overall Sustainable Development of Agriculture/ Horticulture

B.L. Sarswat

Director (C/C) & Executive Director, National Bee Board, Department of Agriculture Cooperation & Farmers Welfare, Ministry of Agriculture and Farmers Welfare Govt. of India

Abstract

Beekeeping is an agro-based activity which is being undertaken by farmers/ landless labours in rural area as an integrated framing practice. It supplements income & employment generation and nutritional intake. Honeybees have been offering services to the society through ensured pollination in crops and providing honey and a variety of other bee hive products viz.; royal jelly, pollen, propolis, bee wax, bee venom, etc. Honey Bees have their vital role in sustaining plants bio-diversity resulting in environmental stability. Mainly five species of honeybees viz.; Apis cerana, A. mellifera, A. florea, A. dorsata, Stingless bees are in used in India. Beekeeping industry has quadruple benefits: 1) providing self employment to rural and forest based population; 2) production of honey, pollen and beeswax, venom, royal jelly, etc., 3) providing employment to rural educated youths in collecting, processing and marketing of bee/beehive-products; and 4) cross-pollination of various agricultural and horticultural crops and improving their quality and increasing yields.

1. Keeping in view the importance of beekeeping, Govt. of India has taken so many initiatives, including efforts of KVIC/State Khadi & Village Industry Boards (SKVIBs), inclusion of beekeeping as a component in National Horticulture Mission (NHM) & Horticulture Mission for Development of Horticulture in North East and Himalayan States (HMNEH)/ Mission for Integrated Development of Horticulture

(MIDH), reconstitution of National Bee Board & designating it as Nodal/National Level Agency etc. Presently, about 30 lakhs bee colonies, with estimated annual production of around 89000 metric tonnes of Honey (2015-16), are maintained in bee hives in the country. The major markets for Indian honey are Germany, USA, UK, Japan, France, Italy, Spain etc.

2. Beekeeping can be initiated with minimal funds and can generate 3.75 lakhs man-days to maintain 10,000 Bee colonies in Bee hive. Income from 100 Bee colonies is around Rs. 2.50-3.00 lakhs per annum. Beekeeping may be adopted as an enterprise by anyone with minimum resources after getting training on the subject.

3. India has vast potential for Beekeeping. Beekeeping industry has great self-help potential for the rural people, tribals, marginal and small farmers, land-less labourers, etc. As per the cropped area under the major insects pollinated crops, about 200 million bee colonies are required in the country to enhance the yield levels of these crops at par with the yield levels of developed countries. It will provide jobs to about 215 lakh persons. Beekeeping may be a source of livelihood for rural and tribal population.

Keywords : *Beekeeping, Cross-pollination, Agro-based activity and Bio-diversity*

A. Introduction

Beekeeping is a very fascinating occupation. Though the honeybees are best known for the honey they produce, their economic role in nature is to pollinate hundreds and thousands of flowering plants and assure setting of seed or fruit.

Sh. Radha Mohan Singh, Hon'ble Minister of Agriculture and Farmers Welfare, Govt. of India addressing participants of State Level Seminar on Beekeeping organized by NBB in Horti. Sangham at Motihari, Bihar in April, 2015.

मधुमक्खियों की मदद से बढ़ेगा फसलों का उत्पादन

इलाहाबाद। खेत के आसपास मधुमक्खी पालन करके फसलों का उत्पादन भी बढ़ाया जा सकता है। मधुमक्खियां रस लेने के लिए एक से दूसरे फूल पर जाती हैं जिससे परागण की क्रिया तेज हो जाती है, जो फसलों के लिए जरूरी होता है। रसायनों के अत्यधिक उपयोग से फसलों के मित्र कीट-पतंगों के विलुप्त होने के कारण उनकी उपयोगिता और बढ़ जाती है। विशेषज्ञों का दावा है कि कृषि क्षेत्र में भी इसका उपयोग बढ़ाकर तथा अन्य उत्पाद बनाकर पालक शहद उत्पादन से इतर 15 गुना तक लाभ बढ़ा सकते हैं। मधुमक्खी के इन गुणों को देखते हुए सरकार ने भी मधुपालन को बढ़ावा देना शुरू कर दिया है। इसी उद्देश्य से कृषि और सहकारिता मंत्रालय के अंतर्गत अलग से नेशनल बी बोर्ड का गठन किया गया है। बी बोर्ड ने माध्वी पालकों के साथ ही अफसरों के लिए भी प्रशिक्षण और जागरूकता कार्यक्रम चलाने का निर्णय लिया है। इसी क्रम में बोर्ड के सहयोग से कुमार ग्रामीण विकास सेवा संस्थान की ओर से पथरचट्टी रामलीला मैदान में चार और पांच मई को कार्यशाला आयोजित की गई। कार्यशाला को संबोधित करने के बाद प्रेसवार्ता में बोर्ड के अधिशासी निदेशक डॉ. आरबी सारस्वत ने बताया कि किसानों को उनके शहद का उचित मूल्य मिल सके, इसके लिए राष्ट्रीय स्तर पर प्रयास किए जा रहे हैं। कार्यशाला में किसान समितियों से जुड़े कई लोग मौजूद रहे।

डॉ. बीएल सारस्वत।

Honey bees thus play very important role in pollinating various agricultural and horticultural crops and increase their yield and improve the quality of produce. Beekeeping supplements income generation to the rural people. Honeybees have been offering services to the society through ensured pollination in cross-pollinated crops as well as by providing honey and a variety of beehive products. It is being increasingly realized that bees could be less expensive input for promoting sustainable and eco friendly agriculture and enhancing crop productivity. The potential benefits, due to bee pollination, in the form of increase in yields of various crops including fruits & vegetables, oilseeds, pulses and others, varies from 5% to 33150%. Studies reveal that the income generated through enhancement in crop yield is much more than the income generated from honey production. Honey Bees have their vital role in sustaining plants bio-diversity resulting in environmental stability.

2. Beekeeping industry has quadruple benefits; 1) providing self employment to rural and forest based population, 2) production of honey, pollen and beeswax, venom and royal jelly, 3) providing employment to rural educated youths in collecting, processing and marketing of bee-products and the most important 4) cross-pollination of various agricultural and horticultural crops and improving their quality and increasing their yields.

B. Beekeeping Industry in India

3. After independence, Govt. of India took policy decision to revive various traditional village industries and an All India Khadi and Village Industries Board (KVIB) was formed in 1954. Through coordinated efforts of well-knit organizations like KVIC, State KVIBs, Beekeepers' Co-operatives, Public Institutions, etc. the beekeeping industry came on the map of village industries of India within two decades. In 1981 an All India Coordinated Research Project on Honeybee Research and Training was launched by ICAR involving Agricultural Universities.

Shri Narender Modi, Hon'ble PM (then CM, Gujarat)
viewing the process of honey extraction.

4. In1994-95, the Ministry of Agriculture, renamed as Ministry of Agriculture and Farmers Welfare, took initiative of launching a Central Sector Scheme entitled "Development of Beekeeping for Improving Crop Productivity", during the VIII Plan. The scheme had various components covering R & D, production & distribution of honey bee colonies, organizing trainings and awareness programmes and support for setting up honey processing plant, etc. A Beekeeping Development Board also functioned under the Chairpersonship of Secretary (A & C) to coordinate the beekeeping activities. The Scheme was approved for continuation during the IX Plan. However, the scheme got subsumed under the Macro Management Scheme, with effect from October, 2000. The focus on beekeeping also got diminished under such an arrangement. The Department facilitated efforts by forming the National Bee Board as a Registered Society under Societies Registration Act through Small Farmers' Agri.-Business Consortium (SFAC) in 2000. With effect from May, 2005, beekeeping has been included as an activity under National Horticulture Mission (NHM) for promoting cross pollination of Horticultural Crops.

5. At present there are about 30 lakhs bee colonies in India, with estimated annual production of around 94,500 (2016-17) metric tonnes of Honey including honey from wild honey bees. India is one of the honey exporting countries. The major markets for Indian honey are Germany, USA, UK, Japan, France, Italy, Spain etc..

6. In India two types of honey viz.; apiary honey (of domesticated bees) and squeezed honey (of wild bees) is produced. *Apis cerana and Apis mellifera* are two types of bees which are being domesticated and kept in hives.

7. In India honey is not used in the form of food as its per capita per year consumption is less than 5.00 gms. But in other countries where it is considered as a food for example, in Germany, per capita honey consumption is more than 2.00 kg. In world, on an average, its per capita consumption is 250-300 gms. whereas in Asia, Japan has the highest per capita honey consumption i.e. about 600 gms.

C. World scenario of beekeeping

8. Millions of honey bee colonies, mostly, Apis mellifera, are maintained all over the world. The world production of honey has bee ranging between 14 to 15 lakh Mt. Tons per year. There are 15 countries in the world which account for the 90 per cent of the world honey production.

China is the only Asian country producing nearly 2.5 lakh metric tonnes of honey. China also produces good quality of beeswax, bee pollen and royal jelly. China is the one of the biggest exporters of honey, beeswax and other bee products. China like India has indigenous *A.cerana* bee colonies and has introduced European *A.mellifera* bees. China have a plan to increase bee colonies in big way during next few decades.

D. Constraints in Beekeeping

9. The The major constraints confronting the development of beekeeping are given as under:

i) Lack of scientific data on the choice of honey bee species for commercial beekeeping and for promoting cross pollination;

ii) Lack of quality nucleus stock of bees;

iii) Lack of infrastructure for producing large volumes of genetically superior queen bees for supply to beekeepers;

iv) Lack of technical knowledge for efficient management of bee colonies for higher honey yield;

v) Lack of infrastructure at grass root level and national level for promoting beekeeping;

vi) Poor quality control for production of honey and other beehive products;

vii) More emphasis for production of honey instead of other bee products such as bees wax, pollen, propolis, bee venom and royal jelly;

viii) Lack of adequate laboratories for disease prevention, control, analysis and testing of quality of bee hive products;

ix) Lack of institutional support for beekeeping in terms of bank loans, etc;

x) Lack of proper pricing policy for honey and those engaged in packaging, processing and storing honey;

xi) Lack of consumer awareness of honey and its products;

xii) This is a non-traditional newly introduced industry;

xiii) Tribals and illiterates from forests and remote rural areas generally practice this industry;

xiv) In this industry there is an interaction of two living materials-honeybees and living plants;

xv) Flowering of plants and secretion of nectar and production of pollen – sole food of honeybees, is influenced by climatic conditions;

xvi) The behaviour and life cycle of honeybees depend completely on climatic and floristic conditions, which vary from place to place;

xvii) De-forestation;

xviii) Indiscriminate use of insecticides, pesticides, weedicides, etc.;

xix) Wild fires;

Officers of NE States learning procedure of honey extraction while participating in training in beekeeping conducted by NBB.

xx) Mono-Cropping culture;

xxi) Pollution of water and air; and

xxii) Global warming & unforeseen changes in climatic conditions.

E. Potential/ Opportunities

10. India has vast potential for Beekeeping. The diversity in flora and fauna provides more opportunities for the development of beekeeping industry. The National Commission on Agriculture had visualized the need for deploying about 150 million Bee colonies for pollinating the agricultural crops in the country. This industry does not need any sophisticated technology, high capital investment or infrastructure. Compared to the potential, not even the fringe of it, in terms of number of honeybee colonies is achieved. There is thus great potential and opportunities for the development of beekeeping industry in India.

11. Beekeeping industry has great self-help potential for the rural people, tribals, marginal and small farmers, land-less labourers, etc. The great potential and opportunities in beekeeping are given as under:

a) As per the cropped area under the major insects pollinated crops, **about 200 million** bee colonies are required in the country to enhance the yield levels of these crops at par with the yield levels of developed countries. It will provide **jobs to about 215 lakh persons.**

b) Honey has great food value and provides cash income;

c) Beeswax which is twice as much costly as honey is in great demand;

d) Other products, viz., bee-collected pollen, propolis, bee-venom and royal jelly are several times costlier than honey and beeswax;

e) Providing bee pollination service to farmers for increasing crop production & quality and productivity of honeybees – A double benefit service;

f) Maintenance of biodiversity by pollination of flowering plants;

g) Apitherapy medicine using bees' products; and

h) Processing and value added products of bee-hive products; etc.

F. Role of Honeybees in increasing production & productivity of Agricultural and Horticultural crops.

12. Till mid-20th century, honeybees were equated with the production of honey and beeswax. But since past 3-4 decades, utilizing honeybees to pollinate large number of agricultural and horticultural crops to increase per acre yield has

become a routine practice in many developed countries. Many commercial beekeepers in America prefer to provide honeybee colonies on rental basis for pollination service rather than to take honey production. According to Agricultural Scientists value of additional yield obtained by pollination service rendered by honeybees alone is about 15-20 times **more** than the value of all the hive products put together (Dr. Kaloo, 2004).

सतना

ग्रेट विक्टर्स के तत्वाधान में नारी जागरण कार्यशाला सम्पन्न

मधु मक्खी फसलों के पैदावार बढ़ाने में मददगार

13. Experiments on effect of bee pollination on various crops were conducted by Central Bee Research and Training Institute (CBRTI) and various Agricultural Universities under All India Coordinated Project on Honeybee Research and Training (ICAR). A cross section of the cross fertile crops, self sterile crops with different degree of self-sterility and even self fertile crops benefited by bee pollination are: **Oilseed** (Mustard, rape seed, toria, lahi, safflower, sunflower, etc.), Orchard Crops (apple varieties, pears, plums, cherry, strawberry, raspberry, persimmon, litchi, citrus varieties, grapes, cucumbers, squashes, melons, alomond, peach, guava, gooseberry, etc.), Legumes (Alfalfa, berseem and other clovers, vetches, broad beans, dwarf beans, Arhar, etc.) and Vegetables (radish, cabbage, turnip, carrot, onion, cauliflower, gourds, etc). The crops-wise details of increase in yield of various crops due to bee pollination are given as under:

Oilseeds	% increase in yields	Legume/ pulses	% increase in yields
Mustard	128.1 to 159.8	Alfalfa	23.4 to 19,733.3
Rai	18.4	Berseem and other Clovers	23.4 to 33,150
Rapeseed	12.8 to 139.3	Vetches	39 to 20,000
Toria	66 to 220	Broad Beans	6.8 to 90.1
Sarson	222	Dwarf beans	2.8 to 20.7
Safflower	4.2 to 114.3	Kidney beans	500 to 600
Linseed	1.7 to 40	Runner beans	20.6 to 1,100
Niger	260.7	Other pulses (Arahar, etc.)	27-30 (RAU)
Sunflower	20 to 3,400	Vegetables for seed/ fruits	% increase in yields
Orchard crops	% increase in yields	Radish	22 to 100
Apple varieties	180 to 6,950	Cabbage	100 to 300
Pears	240 to 6,014	Turnip	100 to 125
Plums	6.7 to 2,739	Carrot	9.1 to 135.4

Contd.

Cherry	56.1 to 1,000	Onion	353.5 to 9,878
Straw-berry	17.4 to 91.9	Brinjal	35-67
Raspberry	291.3 to 462.5	Cucumbers	21.1 to 411
Persimmon	20.8	**Miscellaneous crops**	
Litchi	4,538 to 10,246	American cotton	5 to 20
Citrus varieties	7 to 233.3	Egyptian cotton	16 to 24
Grapes	756.4 to 6,700	Buckwheat	62.5
Squashes	771.4 to 800	Coffee	16.7 to 39. 8
Guava	70-140		
Papaya	22.4-88.9		
Mosambi	36-750		
Orange	471-900		

14. Honey bees have also been identified as major pollinators for spice crops including cardamom, clovers, fennel, coriander, cumin seeds (jeera), fenugreek seeds (Methidana), carom/thymol seeds (ajwain), etc. The bee pollination not only results in higher yields, it may also give a better quality of produce (fruits/ seeds), and the efficient pollination of flowers may also serve to protect the crops against pests.

15. Further the following benefits are also associated with the beekeeping:

(i) **Employment Generation:** Beekeeping has vast potential in employment generation and according to estimates 3,00,000 man-days are generated for maintaining 10,000 bee colonies. Besides, it also creates employment opportunities in appliances and equipment manufacturing sector. It is estimated that around 75,000 man-days are created for manufacturing enough appliances for 10,000 bee colonies.

Regional Seminar on Beekeeping at Barabanki (U.P.)

(ii) **Enhanced Income Generation:** Production and employment are the basic factors to bring enlistment and prosperity. As per economics worked out for beekeeping, the net income from 100 bee colonies varies from Rs. 3,00,000/- to Rs. 5,00,000/- annually, which is directly linked with the prevailing market prices of honey & other beehive products and Beekeeping equipments.

16. It may be remembered that all the bee products are used either as food or in pharmaceutical and cosmetic industries. For this reason, hygienic collection, handling, processing, storage, etc. and maintaining National and International purity standards are of prime importance.

17. It is also worthwhile to mention that beekeeping /honey bees work as an input of agriculture which is essentially required for its development. Therefore, beekeeping/ honey bees should be treated as fifth input for overall development of Agriculture in sustainable manner in the country. It regulates the efficacy of other four inputs used in agriculture, particularly in the crops which need pollinators for pollination support. World wide about 85% of the crops grown need pollinators for pollination and setting of fruits & seeds and honey bees are the best and most important pollinator which can sustain with changes in atmosphere/ climate.

G. National Bee Board (NBB) and its Role

18. NBB is a registered society which was re-constituted under the chairmanship of Secretary (AC&FW) in June, 2006 by DAC&FW, Government of India. The main objective of NBB is overall development of beekeeping by promoting scientific beekeeping in the country to increase the productivity of crops through pollination support and production of honey and other beehive products to increase the income of farmers/beekeepers. NBB is one of the National Level Agencies (NLAs) under MIDH.

19. The main objective of the NBB is overall development of scientific beekeeping in India by popularizing state of art technologies relating to nucleus stock production, capacity building and training of bee breeders and beekeepers, processing, etc. and to increase the productivity of crops through pollination and increasing the honey production for enhancing the income of the beekeepers/ farmers.

20. The funds ear-marked under pollination support component of National Horticulture Mission (NHM) and Technology Mission for Integrated Development of Horticulture in N.E. and Hilly States (TMNE now HMNEH), which have now been merged into Mission for Integrated Development of Horticulture (MIDH), / are used for overall development of beekeeping in the country. The State Departments of Horticulture /State Horticulture Missions (SHMs) are the implementing agencies for these Missions (NHM/HMNEH/ MIDH) at State level. The Board (NBB) which has been recognized as nodal agency to promote scientific beekeeping in the country, has been involved in conducting various capacity building programmes including seminars, trainings,

publications, etc. under NHM/HMNEH/MIDH schemes of DAC&FW as per availability of funds. There is need to make efforts to marry and integrate the beekeeping programme with other programmes of agricultural and rural development at all the levels.

H. Assistance available under MIDH for Promotion & Development of Beekeeping

21. Under MIDH, among others, assistance for following components for promoting Scientific Beekeeping under the component of "Pollination Support through Beekeeping" is available and scheme is being implemented by the State Departments of Horticulture/Agriculture in the field:

S.N	Components	Rates of assistance approved under MIDH (NHM/ HMNEH)
1.	Development and Multiplication of Bee Stock	
i.	Production of nucleus (Pedigree) stock.	Rs.20.00 lakhs/ project
ii.	Production of Bee colonies by Bee Breeders.	40% of cost or Rs.4.00 lakhs/project (whichever is less).
2.	Distribution of 8 frame bee colonies & beehives (50 bee colonies in beehives per beneficiary).	40% of cost or Rs.1600/ per set of bee colony & beehive (whichever is less).
3.	Distribution of bee equipments [a set of one honey extractor of SS (4 frames) & 10 containers (30 kg each) of set / per beneficiary FGP /SS, 1 net & a set of other tools] / unit of 50 bee colonies / beneficiary.	40% of cost or Rs. 8000 per (whichever is less).
4.	Human Resource Dev.(HRD) activities.	
A.	Conf. /Seminar /Workshop:	
	(i) National level	Rs. 5.00 lakh /event
	(ii) State level:	Rs. 3.00 lakh/ event
	(iii) Distt. level	Rs. 2.00 lakh/ event
B.	Trainings – within State Training (WST) & out of State Training (OST)	@ Rs. 1000/- participant/ day & project based (OST)
C.	Exposure Visits – Outside States and Outside India	Project based

I. Beekeeping as an Enterprise, source of Livelihood and benefits

22. Beekeeping industry is source of livelihood for rural poors/ tribals/ forest based population. Benefits of beekeeping are summarized as under

- Unemployed youth can start this business with minimal funds (Rs. 1.00 to 2.00 lakhs)

- Generates 3.75 lakhs mandays to maintain 10,000 Bee colonies in Bee hives;
- Proper utilization of natural resources – nectar & pollen otherwise go waste;
- Different sectors and trades benefit from a strong beekeeping industry;
- Beekeeping encourages ecological awareness;
- Beekeeping helps in increasing National income;
- Income from 100 Bee colonies is around Rs. 2.50-3.00 lakhs per annum;
- May help in doubling farmers income by supplementing/complimenting agriculture/ horticulture;
- Export of honey/beehive products attracts foreign exchange;
- It helps in rural development and promotes small village industry;
- Beekeeping is benign: Beekeeping generates income without destroying habitat;
- Encouraging beekeeping encourages biodiversity.

Hence, beekeeping may be adopted as an enterprise by anyone after getting training on the subject.

Dr. Michael Guido Eich, Expert, Beekeeping, QSI, Bremen, Germany addressing participants of 7 days training programme on scientific beekeeping organized by NBB from 28th November to 4th December, 2011.

23. Presently, the main thrust of NBB is setting up of Integrated Beekeeping Development Centres (IBDCs)/Centres of Excellence (CoEs) on beekeeping, at least one in each State. In these centres (IBDCs), requisite infrastructural facilities for implementing end to end approach for development of scientific beekeeping in the country may be made available at one place. Centres will

help the beekeepers/farmers of the area in adopting scientific beekeeping and encourage/promote scientific beekeeping in integrated manner in the Country. 3 IBDCs have been commissioned/ approved by 2015-16 and 7 have been approved/ sanctioned during financial year (2016-17). NBB has plan to develop at least one IBDC in each State in the country. Other projects/ schemes/ new initiatives taken by NBB for promoting scientific beekeeping are:

i) Pilot project for Integrated Development of Scientific Beekeeping for enhancing crop productivity & income of farmers/beekeepers and generating employment in rural areas under RKVY Scheme during 2016-17 for its implementation in 8 States, viz., Assam, Bihar, Gujarat, Karnataka, Maharashtra, HP, Rajasthan and UP (2 Districts in each).

ii) 2 Projects: (i) Model-I:- For adoption of Scientific Beekeeping by farmers as an enterprise and (ii) Model-II:- Providing pollination supports by beekeepers/beekeeping entrepreneurs on custom hiring basis submitted to various Divisions of DAC&FW.

iii) All concerned Divisions/ Departments/ Organisations have been requested to treat beekeeping/honey bees as 5th input in Agriculture/Horticulture.

iv) Registration of beekeepers/societies/firms/companies involved in beekeeping.

v) National Dairy Development Board (NDDB), Gujarat Cooperative Milk Marketing Federation (GCMMF)/ AMUL/ Milk Cooperatives/Federations convinced for their involvement in beekeeping, collection and marketing of honey. In pursuance of that the Safal brand of honey has been launched by Mother Dairy under overall guidance of NBB.

vi) The representatives of Banas Dairy Union, Palanpur, Banaskantha were convinced and trained for promoting beekeeping among their members/ milk producers/societies. This resulted release of Banas Brand of Honey launched by Hon'ble Prime Minister on 10th December, 2016 at Palanpur, Gujarat.

vii) Participating in various Horti/Agro Expos and disseminating information on scientific beekeeping. In the year 2016, NBB participated in Krishi Unnati Mela, 2016 and demonstrated world class technologies on scientific beekeeping for which NBB was awarded (First Award).

24. Main issues to be addressed are strengthening of National Bee Board, setting up of State Bee Boards/Missions/IBDCs; production of quality germplasm & nucleus stock of honey bees; indiscriminate use of pesticides in crops; quality standards for honey & other beehive products by BIS/ FSSAI, etc.; disease diagnostic labs & bee products quality analysis labs ; exemptions from various

taxes (GST) for Beekeeping/beekeepers; treating beekeepers as farmers in all respects for compensation, etc. in the event of damage of bee colonies and accidental insurance coverage on subsidized rates, insurance of bee colonies on subsidized rates, etc.

9

Horticulture Based Integrated Farming Systems: Models for Small and Marginal Farmers

Sanjeev Kumar and Shivani

ICAR Research Complex for Eastern Region, Patna-800014, Bihar

Introduction

Farming and agriculture are respectively defined as the practice of cultivating the land or raising stock and the production of agricultural goods through the growing of plants and the raising of domesticated animals. In the developing countries frantic efforts have been made through research and enlightenment campaigns to encourage farming and thus ensure significant increases in agricultural production in order to feed and sustain the population that is increasing at geometric proportion. At present, the farmers concentrate mainly on crop production which is subjected to a high degree of uncertainty in income and employment to the farmers. In this context, it is imperative to evolve suitable strategy for augmenting the income of a farm.

Integrated farming (or integrated agriculture) is a commonly and broadly used word to explain a more integrated approach to farming as compared to existing monoculture approaches. It refers to agricultural systems that integrate livestock and crop production. Integrated farming system has revolutionized conventional farming of livestock, aquaculture, horticulture, agro-industry and allied activities. It could be crop-fish integration, livestock- fish integration, crop-fish-livestock integration or combinations of crop, livestock, fish and other enterprises

The integrated farming system approach introduces a change in the farming techniques for maximum production in the cropping pattern and takes care of optimal utilization of resources. The farm wastes are better recycled for productive purposes in the integrated system. A judicious mix of agricultural enterprises like dairy, poultry, piggery, fishery, sericulture etc. suited to the given agro-climatic conditions and socio-economic status of the farmers would bring prosperity in the farming. An integrated farming system allows us to use some of the advantages of nature, and ecology, as opposed to relying on chemistry to solve all our production issues.

Throughout India, our present agriculture concerns are multifaceted and multidimensional. Some notable challenges include pressure of the population on land, skewed distribution of operational holdings, land degradation, imbalance use of water and nutrients, low productivity, low level of mechanization, low fertilizer consumption, climate change impact, non-remunerative prices and postharvest losses. Nevertheless, the eastern region of the country holds promise for a second Green Revolution through holistic management of land, water, crops, biomass, horticultural, livestock, fishery and human resources. The region has about 75% marginal farmers. There is tremendous pressure on land to sustain the ever increasing human and bovine population. Small and fragmented landholdings limit, by and large, the adoption of latest farming practices. The coastal areas are also vulnerable to sea-water intrusion and cyclones. Further, the wetlands, which occupy an area of 4.05 million ha, also need restoration for sustainable food production and conservation of biodiversity. The region also has about 162.64 million underprivileged people compared to 360 million at the national level, which need immediate attention so as to improve upon their livelihood options. The region also has about 36 million schedule tribe (ST) population (which is 1/3 of the total ST population at the national level) and, therefore, promising tribal agriculture practices needs to be replicated in similar agro-climatic zones for better livelihood options of ST population.

The challenges ahead in agriculture sector consist of development of (i) sustainable agriculture, (ii) food and nutritional security, (iii) climatic vulnerability and its impact on crop productivity, (iv) generation and transfer of technology, (v) Enhancement in resource use efficiency and water productivity (vi) provision of incentives for agriculture, (vii) promotion of investments in agriculture, (viii) strengthening of institutional infrastructure, (ix) better risk management, (x) corporate agriculture, (xi) peri - urban agriculture, and (xii) tapping the potential of eastern region and introduction of management reforms.

In order to meet out aforementioned challenges, technology integration, scaling up and framing of demand driven productivity enhancing research agenda in a network mode, using both conventional and frontier technologies ensuring scientific management of natural resources and production sustainability is essentially required since agricultureis, by and large, diverse and risk prone. To feed the Indian ever-growing population it is demand of the time to switch over from traditional farming to Integrated farming system mode. Some of the opportunities which needs special attention towards sustainable production are - System Mode Production, Intensive Integrated Farming Systems (IIFS): a sustainable alternative for the benefit of small scale farmers, Synergistic role of livestock in farming systems, Integrated fish farming, Conservation and sustainable use of Wetlands, Agroforestry interventions, Conservation agriculture, Harnessing green energy etc. and these all opportunities can be carried out together under one umbrella of Integrated farming system (IFS).

Most of the Indian farmers are so poor that they can't afford even two times meal for their family. Under such circumstances it will be injustice to talk about nutritional security for such farmers in present situation. These poor farmers with small land holdings concentrate only on crop production, mainly cereals with high risks of flood and drought. Due to failure of monsoon and small size of holdings, they hardly get sufficient income to sustain their family. To solve the problems of small resource poor farmers, diverse and risk prone environments has led to the development of a more holistic, resource based, client oriented and interacting approach, popularly known as Integrated Farming System (IFS). These small and marginal farmers can take a suitable crop along with horticulture, animals, fisheries and other components in integrated manner that would minimize risks and provide additional income and employment from the same piece of land per unit time, apart from improving soil fertility over a period of time. Beside this fact it is assumed that if farmers will cultivate their land in IFS mode then they will harvest quality food items i.e. cereals/fruits/milk/eggs/meat etc. in one basket and be able to fulfill the nutritional requirement of their family at the sametime.

Integrated Farming System: Concepts

Farming System research is an approach to agricultural research and development that view the whole farm as a system and focus on 1) the interdependencies between the components under the control of members of the household and 2) how these components interact with each other in respect of physical, biological and socioeconomic factors not under the household's control. Indian economy is predominantly rural and agriculture oriented where the marginal and small farmers constitute 76.2 % of farming community. Due to failure of monsoon, the farmers are forced to judicious mix up of agricultural enterprises like dairy, poultry, pigeon, fishery, sericulture, apiculture etc., suited to their agro-climatic and socio- economic condition.

Why integrated Farming System

It enables to identify the major constraints in increasing farm productivity.

- It enables to provide technological intervention options for improving farming systems at a given resource base through farmer participatory research.

- To understand the physical and socio-economic environment within which agricultural production takes place
- To gain an understand of the farmer in terms of his/her skills, constraints, preferences and aspirations
- To comprehend and evaluate existing important farming systems, in particular, the practice and performance of these systems.

Advantages of Integrated farming Systems

In an integrated system, livestock and crops are produced within a coordinated framework. The waste products of one component serve as a resource for the other. For example, manure is used to enhance crop production; crop residues and by-products feed the animals, supplementing often inadequate feed supplies, thus contributing to improved animal nutrition and productivity

- The result of this cyclical combination is the mixed farming system, which exists in many forms and represents the largest category of livestock systems in the world in terms of animal numbers, productivity and number of people it serves.
- Integration of allied activities will result in the availability of nutritious food enriched with protein, carbohydrate, fat, minerals and vitamins.
- Integrated farming will help in environmental protection through effective recycling of waste from animal activities like piggery, poultry and other birds rearing.
- Animals play key and multiple roles in the functioning of the farm, and not only because they provide livestock products (meat, milk, eggs, wool, andhides) orcanbe converted into prompt cash in times of need. Animals transform plant energy into useful work: animal power is used for ploughing, transport and in activities such as milling, logging, road construction, marketing, and water lifting for irrigation.

Scope for Integrated Farming System

Farming enterprises include crop, livestock, poultry, fish, tree crops, plantation crops, etc. A combination of one or more enterprises with cropping, when carefully chosen, planned and executed, gives greater dividends than a single enterprise, especially for small and marginal farmers. Farm as a unit is to be considered and planned for effective integration of the enterprises to be combined with crop production activity. Integration of farm enterprises to be combined on many factors such as:

1.Soil and climatic features of the selected area.

2. Availability of resources, land, labour and capital.
3. Present level of utilization of resources.
4. Economics of proposed integrated farming system.
5. Managerial skill of the farmer

In the context of India, there are a number of situations and conditions that can be alleviated by an Integrated Farming System. The following situations are ideal for the introduction of Integrated Farming Systems:

- The farmer wishes to improve the soil quality
- The farm household is struggling to buy food or below the poverty line
- Water is stored on-farm in ponds or river-charged overflow areas
- Fertilizers are expensive or the recommended blend is unavailable
- Soil salinity has increased as a result of inorganic fertilizer use
- The farmer is seeking to maximize profits on existing holding
- The farm is being eroded by wind or water
- The farmer is looking to reduce chemical control methods
- The farmer wants to reduce pollution or waste disposal costs

Integration of Enterprises

Since Integrated Farming System (IFS) is an interrelated complex matrix of soil, water, plant, animal and environment and their interaction with each other enable the system more viable and profitable over arable farming system and leads to produce the quality food. The

income obtained from crops may hardly be sufficient to sustain the farm family throughout the year. Assured regular cash flow is possible when the crop is combined with other enterprises. Judicious combination of enterprises, keeping in view of the environmental conditions of a locality will pay greater dividends. At the same time, it will also promote effective recycling of residues/wastes (Kumar *et al.*, 2012).

To strengthen the food chain, it is essential to eliminate nutritional disorder which has been realized on account of appearing deficiencies of mineral nutrients and vitamins in food being consumed. Horticultural and vegetable crops can provide 2-3 times more energy production than cereal crops on the same piece of land and will also ensure the nutritional security and income sustainability in the existing system. (Gill *et al.*, 2009). Similarly inclusion of bee- keeping, fishery, mushroom cultivation, bird rearing, goatry, livestock on account of space conservative also provide additional high energy without affecting production of food grains. Integration of these enterprises helps the production, consumption and decomposition in a realistic manner in an ecosystem. It is pre- requisite in farming system to ensure the efficient recycling of resources particularly crop residues and animal wastes, because 70-80 percent of the micronutrients remain in the biomass and animal wastes (Inman *et al.*, 2005).

It will be wise to select the enterprise by keeping the location specificity in mind, means on the basis of environmental condition of the area, land topography, soil and ecosystem, market and processing facility, socio- economical condition, risk bearing capacity, knowledge about the selected enterprise and investment capacity of the individual as farming models are highly location specific and it varies from place to place and even farmer to farmer in the same area. Therefore, proper attention is required while selecting an enterprise to integrate into the system so that the farming model would be profitable and sustainable in all respect.

Integrated farming system models for different Ecosystems

IFS models for dry Regions

Crop + dairy + biogas + silviculture

The major cropping systems followed under irrigated uplands are Sorghum - cotton - maize, Ragi - cotton - maize and Ragi - cotton -sorghum. The various components integrated were dairy, biogas, spawn and mushroom production, apiary, homesteadgarden, fish rearing and silviculture. Integrated farming system involving crop, dairy, biogas and silviculture was taken from 1987-92, The normal cropping pattern followed in Coimbatore is ragi-cotton-sorghum. In the integrated approach, the same cropping pattern was slightly modified by inclusion of intercropping such as green gramin cotton, cowpea in sorghum and sunflower as border crop in ragi (0.75 ha).The perennial grass fodder (0.15 ha) and legume fodder Lucerne (0.05 ha) were also raised in the holding.

Three Jersey crossbred milch cows with 2 calves were included under dairy component. For effective recycling of farm and animal waste, a biogas unit of 2 m^ capacity was installed for the production of fuel, light and enriched manure. Sixty kgof cow dung expected out of 5 animals is sufficient enough to produce 2 $m^3$of gas everyday, which is equivalent to 1.5 litres kerosene. Two hundred numbers of subabul tress were planted all along the boundary of the field for fodder and timber production. The results of the study revealed mat the entire system produced a net income of Rs. 20702/ha/yr. This system also facilitated effective recycling of farm and animal waste, improved farm employment opportunities and continuous flow of income to farm throughout the year.

Crop + dairy + biogas + mushroom + fish

In this system components of integrated farming system involved were crop + dairy, crop + dairy + biogas + fish, crop + dairy + biogas + mushroom + fish and crop + fish. Experimental results on farming systems for' irrigated uplands revealed that integration of cropping with components like dairy, biogas and mushroom or fish resulted in higher productivity than cropping alone. During 1999-2000, integration of 0.32 ha each of sorghum + red gram - sunflower + coriander, maize + fodder cowpea - cotton + coriander and perennial fodder CO- grass + legume fodder (Lucerne) with dairy (6 cows + 4 calves), biogas (2 m capacity), mushroom (2 kg/day) + spawn production (10 bottles/day) and fish (0.008 ha pond with 80 fingerlings fed by cattle shed washing} yielded higher sorghum productivity than cropping alone with sorghum - cotton (0.50 ha) *and* maize - cotton (0.50 ha) cropping systems. During 2000-2001, integration of 0.32 ha each of sorghum + red gram - sunflower + coriander, maize + fodder cowpea - cotton+ coriander and perennial fodder CO_3 grass + legume fodder (Lucerne) with dairy (6 cows + 4 calves), biogas (2 nr[1] capacity') and mushroom (2 kg/day) + spawn production (10 bottles/day) recorded higher productivity than the cropping alone with sorghum - cotton (0.50 ha) and maize - cotton (0.50 ha) cropping systems.

Cropping + dairy + biogas + mushroom + fish (cattle shed washing) integration earned the highest gross and net returns with better per day returns. This farming system registered the benefit - cost ratio of 2.22 during "1999-2000. Cropping + dairy + biogas + fish + mushroom integration recorded the highest gross, net and per day returns. It also registeredthehighestbenefitcostratioof2.41during2000-2001(Sivamurugan,2001).

Crop + goat

An integrated farming system study involving grain crop, fodder crop, fodder trees, perennial grasses and goat rearing in an area of one hectare of rainfed land was carried out at Coimbatore from 1987-1992. Generally, in dryland farmers raise only fodder sorghum yielding, to 10 tonnes of fodder/ha depending upon the rainfall. In the integrated approach, the cropping pattern was modified by including both sorghum grain crop (0.2 ha) sorghum fodder crop (0.2 ha). To meet *the* fodder requirement of goats, 0.2 ha of land was raised with subabul and *Cenchrusciliaris*, a posture grass as an intercrop. Trees like *Acacia Senegal* and *Prosopiscineraria* were raised in 0.20 ha. The goat unit comprised or Tellicherygoats 20 eves and 1buck.

The results revealed that through short duration field crops and perennial crops, the feed requirement for one productive unit consisting of 20 eves and one buck for all the 365 days could be met. After 5 years, the perennial fodder trees could bear the shock and would yield sufficient quantum of loppings to supplement the feed requirement of 21adultsalongwithmillets, legumes and perennial grass linked. The Tell cherry go at will be a good breed to build up body weight for every unit of feed secured through different sources under rainfed condition. It is a dual-purpose animal, where it gives 80 to 100 ml of milk after satisfying the full requirement of dependent kids on it every day. Twenty productive females could give 45 kids per annum. Each kid at the time of weaning will weigh around 12 kg. Moreover the unit of 21 animals with different-stages of kids under deep litter system would give 11.2 t of valuable manure. This, when applied to the soil,will not only an excellent source of primary, secondary and micronutrients for the crops but also absorb more moisture, retain in the soil and releases to the crop appropriately for better yield.

The net income from the farming system was Rs.5671/ha/year and that of control Rs. 1919/ha/year. Out of die total income from the integrated farming systems, 59 per cent was from goat rearing. The additional net income realized from integrated farming system was Rs.3752/ha/year as compared to cropping alone.

Crop + tree + goat

Another integrated farming system model involving crop + tree + goat was taken from March 1999 to February 2001. Conventional cropping system with sole sorghum was compared with *Allan thus excelsa*+ crop + goat, *Ceiba pan tan dra*+ crop + goat and *Emblica officinal is* + crop + goat to identify the suitable component linkage. The cropping systems included were grain sorghum + cowpea, fodder sorghum + cowpea and Cenchrus glaucus each in 0.33 ha in integrated farming systems and the remaining 0.01 ha was allotted to the goat component. One unit of Tellichery goats, consisting five females and one male was included in the system. Three tree species *viz., A. excelsa, C. pentandra*and £. *officinalis*were evaluated for their performance under dryland situation to evaluate the suitable in situ moisture conservation practices and nitrogen management on yield of crops in the integrated farming system. Experimental results on integrated farming system revealed that (i) integration of sorghum + cowpea (grain), sorghum + cowpea (fodder) and C. *glaucus* each in 0.33 ha intercropped in E. *officinalis* with Tellichery goat component (5+1) in 0.01 haresulted in higher productivity, economic returns and provided better employment opportunity and improved soil fertility than raising sole sorghum alone (ii) growth of *E. officinalis* was better as compared to other trees under vertisol dryland situation, (iii) coir pith mulching and pitcher irrigation increased the tree seedling growth than the control, (iv) tied ridges conserved more moisture and improved the productivity of the crops, (v) application of 50 per cent N through fertilizer and 50 per cent N through goat: manure increased the productivity, enhanced the soil fertility and provided better opportunity for recycling of manure to the crops (Radhamani, 2001).

Cropping + fish + poultry

An experiment was conducted during 1992-1994 in rice based farming system as a demonstration trial at Aduthurai. The components were cropping, fish culture and poultry. An area of 0.40 ha was selected "for the farming system study, considering the small and marginal farmers of the 'state. Conventional cropping as practiced by farmers was taken up in an area of 0.96 hectare. In the fish pond with 400 fingeriings belonging to the species viz., *Catla (Catlacatla)* (200) *Roliu (Labeorohita)* (100), *Marignl (Cirrbinusmrigale)* (100) were stocked. Accordingly the stocking density of fingerlings works to 10,000 ha of ponded water without artificial feeding.

A poultry shed was erected over the fish pond at one corner. Bottom of poultry shed was provided with wire mesh (3 mm x 3mm) to facilitate tree failing or poultry droppings into the pond. Total of 100 Bapkok layer birds were kept in the poultry shed. The birds started laying eggs around 22 weeks and were culled at the age of 72 weeks when the egg productions become uneconomical.

The economics worked out for the system as a whole was Rs. 28.983, in which cropping system contributed Rs. 23,709, Poultry and Fisheries contributed additional income of Rs.5,274. Poultry droppings added to the fish pond as feed was 3 tonnes/year (l00g/birds). Mean number of egg production was 262/*year/bird.* In the case of fish pond (0.04) yield recorded was 234 kg, Of the income obtained from the integrated farming system, 78% was from cropping system and poultry cum fisheries generated additional income and employment (Porpavai *et at.,* 2000).

Cropping+duck+fish culture

Integrated farming system with duck-cum-fish culture as a component was tried during the year 1989. Two farm holdings each with the size of one ha were selected for conductingthe study. In one holding, conventional cropping aspracticedby farmers was followed. In another one hectare, cropping was practiced in an area of 0.973 ha and an area of 0.027 ha was allotted for duck-cum-fish culture. Economics of IFS was compared with existing cropping systems. Net income of Rs. 13790/-was obtained from existing cropping system (*Kunivai-thaladi* rice - pulse) and a net income of Rs. 22676/- was obtained from the modified cropping system (*Kurtwai- thaladi* rice - cotton and maize) withan area of 0.973 ha allotted for, cropping. The additional profit from modified cropping alone was Rs.8886/-. From duck-cum-fish culture as a component in mixed farming system, a net profit of Rs. 1441/- was obtained from an area of 0.027 ha. Totally an additional income of Rs. 1032/- was obtained from the mixed farming system over existing cropping system (Ganesan,1989).

Cropping+milch animals

Another comparative study with rice-rice-pulse as conventional cropping system (CCS) and rice-rice-blackgram/ Cotton and NB21 fodder grass and three milch cows as Integrated Farming System were tried. The results revealed that a net return of Rs. 8422/- was obtained from CCS and Rs.10913/- was obtained from IFS. Net return from dairy enterprise was Rs. 8987/-. The additional net income of Rs. 11478/- was obtained from the dairy based mixed farming over the existing farmer's method of cropping (Rajkumar1988). A study by integrating goat (5+1) with crop activity gave a net income of Rs. 25, 400/ha/ year. Additional income over conventional cropping system was Rs. 11,932/ha/year. Similarly in a study, where, sericulture unit was linked with crop activity, the net income was Rs. 22,426/ha/year. Thus, the farmer could get an additional income to the tune of Rs. 9, 951/ ha/year (Kuppusamy, 1989). In yet other study, where, edible mushroom cultivation was linked with crop activity, the net income was Rs. 24,685/ha/year. Integration of edible mushroom provided additional income to the tune of Rs. 12,977/ha/year.

Horticulture + dairy + goat

Integrated farming system experiment was carried out during 1992-96 with multiple objectives of achieving better utilization of available resources for obtaining maximum returns by integrating crop component with dairy and goat unit, to recycle farm and livestock wastes effectively and providing gainful employment throughout the year and to assure stability in production and return. An area of one hectare was selected for this integrated farming system (IFS) study, Conventional cropping system (CCS) as practiced by the farmers was taken up in an area of 0.2 ha for comparison. Tell cherry goats (5 female + 1 male) at the age of 4 to 5 months were chosen for the study. The goats were fed with 4 kg of green fodder and 1 kg of legume/ adult/ day. Out of the total incomeobtained from the IFS, 66 per cent was from crop component and 17 per cent from goat unit. The cost-benefit ratio from crops, dairy and goat units were 2.2, 1.9 and 2.3 respectively. This clearly indicates the feasibility of rearing goats as complimentary enterprise for small farmers of Cauvery delta zone to get additional economic returns. Hence, for the regular flow of income, additional employment generation and also effective recycling of wastes, dairy and goat units would be the ideal complementary enterprises with crop component for the Cauvery delta zone farmers (Sheik Dawood *et al.,*1996).

IFS models for irrigated ecosystem

Representing integrated IFS models 1. One acre IFS model (for mid- land situation) and 2. Two acre IFS model (for lowland situation) were developed at ICAR Research Complex for Eastern Region, Patna through which Income of the farmers' from the same piece of land was enhanced by 3-4 times (Sanjeev *et al.*, 2012). It lies under middle Gangetic plains of sub- humid region at 53.5 m-ams altitude. The average minimum temperature varies from 6°C – 14°C and average maximum temperature between 20-22°C during winter season while average minimum and maximum temperature during summer varies from 22-30°C and 35- 45°C, respectively. Average rainfall of the place varies from 1100-1200 mm in the normal monsoon.

IFS models at a Glance

One acre IFS model	Two acre IFS model
Main enterprises: Crop (Hort. + cereals) + Goat + Poultry Allied enterprises: mushroom, vermicomposting rmicomposting	Main enterprises: Crop (Hort. + cereals) + Livestock (2 no.) + Fishery Allied enterprises: Duckery, composting,

A view of developed IFS model at ICAR-RCER, Patna

Land allocation to different components under two acre IFS model

1) **Cereal crops** (50 % area)

   Kharif: Rice

   Rabi: Wheat/Maize Gram/Mustard

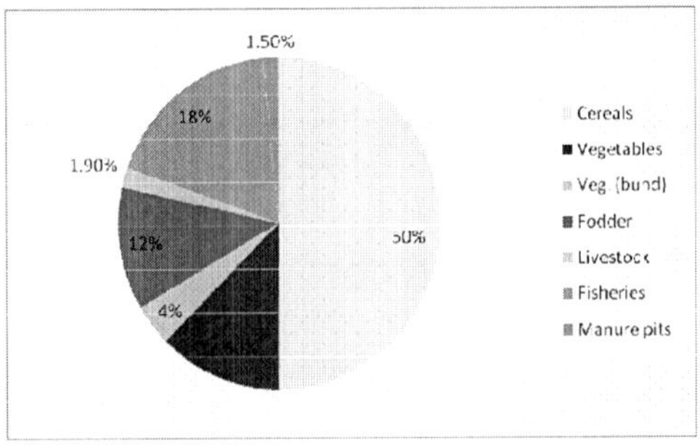

2) **Horticultural crops** (Fruits+vegetables): 12.5% area

Vegetables

Kharif: Cucurbits/Brinjal/Okra

Summer: Onion/Brinjal/Cowpea/Okra/Bitter gourd/Cucumber etc.

Rabi: Tomato, Cabbage, Brocelli, French bean

Fruits

Papaya (On pond's dike and field bunds) Banana (On pond's dike)

Lemon (On pond's dike and Horticultural block)

Guava (On pond's dike and Horticultural block)

3) **Boundary plantations** (4 %area)

All around the fields, drumsticks and dhaincha plants were planted to provide fodder to animals and seeds for green manure crop. On field funds, fencing was done and cucurbits, pigeon pea and soybean crops are being raised for maximum utilization of land and to provide protein supplements to farm families.

4) All around the field bunds cucurbits or seasonal vegetables having lesser water requirement may be raised by making wire fences.

5) Fish + Duck integration (17.8 %area)

a. Mix carp culture: Rohu (20% as column feeder), Catla (30% as surface feeder), Mrigal/common carp (50% as bottomfeeder)

b. Duck: For 1000m^2 water area 40- 45 number of ducks are found sufficient.

Khakhi Campbell breed of duck is right choice for this area (Dual purpose). A thatched hut of 10 X 15' size is optimum for 40 ducks above the water or on the pond's dike.

Fruit+Vegetables plantation around the pond dykes

6) **Livestock** (1.80 % area)+ Bio- gasunit

A size of 2 adult cows + 2 calves is optimum for two acre land in respect of FYM requirement for the fields and fodder requirement for the livestock. A thatched hut of 20' X 30' with sufficient paddock space is sufficient for above no. of animals. The Cow shed was connected with the pond with a drainage channel so that urine and water can move into the pond. A storage hut for storing of animal feed was also made near the animal shed. A bio- gas unit of $2m^3$ capacity was also constructed under livestock area for production of bio-gas for energy and slurry to making vermicompost to the crops. It was found that for $2m^3$ capacity of bio-gas unit, by product obtained from two adult cows are optimum.

7) **Fodder production** (12.5 % area)

For feeding of 2 cows and 2 calves 1000 m^2 land is sufficient if year round fodder production is carried out. In addition to green fodder, straw, leaves, stems of different cereals and vegetables can be also used as animal feed.

Kharif: M.P. Chari/Sudan grass/ Napier/Maize

Summer: Boro/Lobia/Maize/Sudan grass **Rabi:** Berseem/Oat/Maizeetc.

8) **Spices**: In the sheds or where light intensity is less like orchards, spaces between the huts etc. turmeric, ginger or guinea grass are being taken.

9) **FYM/vermicomposting pits**: (1.4 %area)

Optimal sizes pits (9' X 3') for preparation of FYM 93 pits) and Vermicompost (4 pits) has been made. Sizes may depends upon land available near the livestock shed so that required raw materials for making manures should be made available nearby for convenience and to avoid transportation charges.

Note: Cattle shed should be always constructed away from birds to avoid attack of any transmissible or contagious diseases to animals or vice-versa.

Land allocation to different components under one acre IFS model

1. **Cereal crops** (50 %area)

   Cereals

   Kharif: Rice

   Rabi: Wheat/Maize/ Lentil/Til

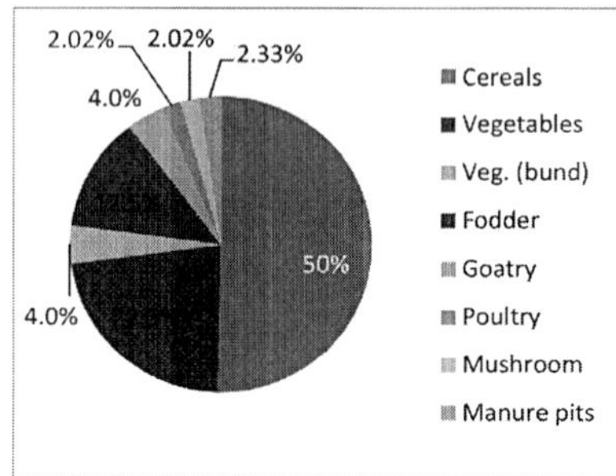

2. **Horticultural crops** (Fruits + vegetables): 22.5 %area

   Vegetables

   Kharif: Cucurbits/Brinjal/Okra

   Summer: Brinjal/Cowpea/Okra/ Bitter gourd/Cucumberetc.

   Rabi: Tomato, Cauliflower, spinach

   Fruits

   Banana (On field bund) Lemon (In Horticultural block) Guava (In Horticulturalblock)

3. **Boundary plantations** (4 %area)

   All around the fields, Karaunda, drumsticks and dhaincha plants were planted to provide fodder to animals and seeds for green manure crop. On field funds, fencing was done and cucurbits, pigeon pea and soybean crops are being raised for maximum utilization of land and to provide protein supplements to farm families.

4) **Livestock (Goat)**: 2.5% area

   A size of 20 female got + 1 buck is optimum for one acre land in respect of manure requirement for the fields and fodder requirement for the livestock. A thatched hut of 20’ X 30’ with sufficient fenced paddock space (to move the goats freely as goats have to kept on stall feeding) is sufficient for above no. of animals. The goat shed was made airy and sunny. A storage hut for storing of animal feed should was also made near the animal shed. Black Bengal breed of goats are found suitable for this region.

5) **Poultry** (200birds)

   200 birds (broiler) are deing reared in an area of 225 sq. ft. by making a thatched hut. All around the that chedhut's walls, wire meshing has been done at the innerwalls to protect the birds from predators and hunting animals. The hut was made airy and proper arrangement of bulbs was made before rearing the chicks.

6) **Mushroom**

   Year round mushroom production is being done in an area of 25 X 20' by making a thatched hut for optimum return. In this shed about 200 mushroom bags are being kept at a time by making bamboo shelves. Selection of the mushroom strains is done on the basis of climate, temperature and humidity in the atmosphere as:

   March – September: straw/paddy/milky mushroom October- February: Oyster/ Button mushroom

7) **Fodder production** (12.5 %area)

   For feeding of 20 + 1units of goat an area of $600m^2$ is sufficient if year round fodder production is carried out. In addition to green fodder, dry husks, leaves, stems of different cereals and vegetables are also being used as feed.

   Kharif: M.P. Chari/Sudangrass

   Summer: Boro/Lobia/Maize/Guneagrass

   Rabi: Berseem/Oat/Maize etc.

8) **Spices:** In the sheds or where light intensity is less like orchards, spaces between the huts etc. turmeric, ginger or guinea grass can betaken.

9) **Compost pits/ vermicomposting pits** (1.4 %area)

   Optimal sizes pits for preparation of goat manure and Vermicompost should be made depending upon land available near goat shed so that required raw materials for making manures should be made available nearby field and livestock.

Note: Goat shed should be always constructed away from poultry shed to avoid attack of any transmissible or contagious diseases to animals.

Income and expenditure from IFS model

The farming system models were started in the year 2010, and on the basis of five years data, now it can be analysed that by integrating Crop + Livestock + Fish/ duck in two acre area of land, a net income of Rs. 1,37,209/- can be achieved with a B:C ratio of 1.8 which is about 4 times higher over rice- wheat cropping system (Table 1) with an additional income equivalent to 118 Kg urea, 247 kg SSP and 71.2 kg SSP as due to nutrient recycling within the system 56.5 kh N, 39.6 kg P and 42.7 kg K were added to the soil which will be utilized by the next crop Table 2) . To start up with all these components an initial investment of Rs. 2, 05,000/- may be required.

Table 1: Estab. Cost, Expenditures and Net income for one acre IFS model

S.No.	Components	Estb. Cost (Rs.)	Recurring expd./ann. (Rs.)	NI at 5th year (Rs.)
1.	Crop (0.2 ha)	-	14,062	14,112
2.	Horticulture (0.09 ha)	5000	10,946	12,843
3.	Fodder	—	10,175	4,165
3.	Goat (20 + 1)(0.018 ha)	65,220	34,632	18,225
4.	Mushroom (0.003 ha)	9,000	6,200	3,461
5.	Poultry (700 chicks)(0.0015 ha)	15,000	64,920	24,282
6.	Crop waste/V.c/FYM pits	8000	3,287	2,163
	Total			

Contributions to income from different enterprises play an important role while selection of components for a particular IFS model development. While selecting the components one should be enough careful about his needs, technical knowledge about the component, water, land and labour availability, transportation and marketing facilities etc. to get maximum profit. In the developed two acre IFS model, Fish integration has resulted in maximum contribution to income and was followed by Horticulture and dairy components while maximum contribution to income was gained towards poultry and was followed by goatry and horticulture under one acre IFS model, respectively (Fig.1).

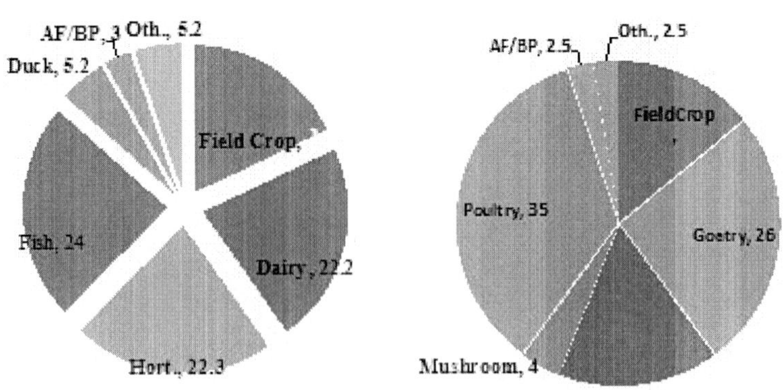

Two acre IFS mode **Two acre IFS mode**

Fig.1. Percent contribution to income from different components under one and two acre IFS models

Nutrient recycling

Nutrient recycling within the system is prerequisite for development or integration of any component in the IFS model. Priorities should be given to those components whose by-product can be recycled within the system or can be reused as input for another component to increase nutrient use efficiency on one hand and also for decreasing the cost of cultivation and addition of organic forms to the system for its sustainability. Under two acre IFS model, 13.8t of cow dung from two cows, 11.3 t of vegetable wastes and 1.21 t of duck dropping were produced and were recycled within the system which added an amount of Rs. 4,826/year to the income. Likewise, 2.5t of goat manure, 6.62t of vegetable wastes, 1.78 t of poultry droppings and 4.64t of rice/maize/lentil straws were recycled within the system which contributed Rs. 3,175 to the income and added 44.0 kg N, 29.5 kg P and 31.2 kg K in the soil which was equivalent to 93.0kg urea, 184.0 kg SSP and 52.0 Kg MOP (Table 4). In addition to these nutrients an ample quantity of micronutrients were also added to the soil upon nutrient recycling. A nutrient flow diagram for the purpose of nutrient recycling in an IFS model has been represented as Fig.2.

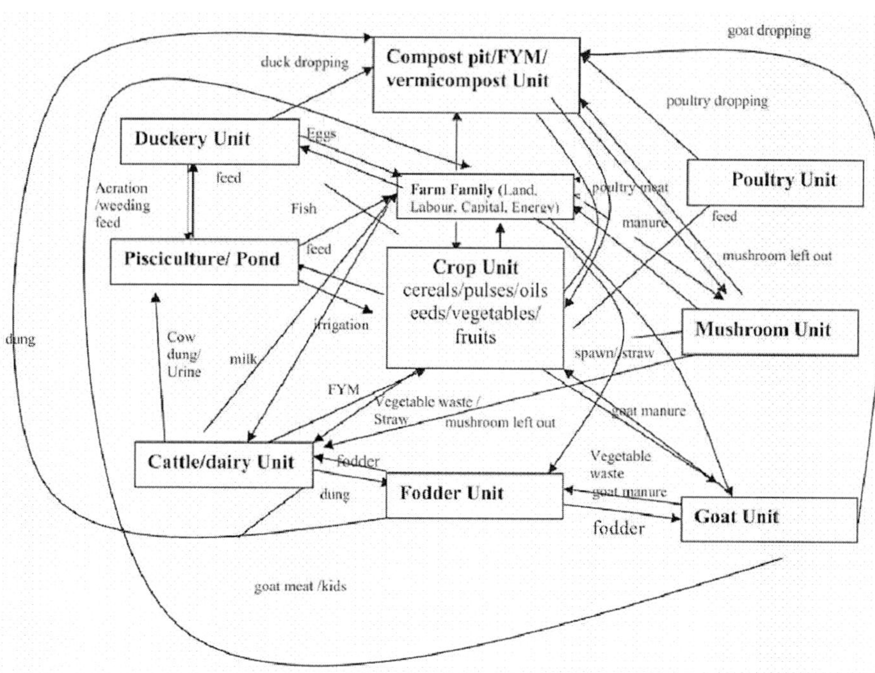

Fig. 2: Input and output flow diagram under IFS model

Table 2: Recycling of farm waste and gain/saving of nutrients through 2 acre IFS

S. No	Farm waste	Quantity produced (t)	Production/use pattern(t)	Nutrient gain (kg)	Total Nutrient Gain from recycling	Saving (Rs.)	Fert. equivalent (kg)
1.	Cow dung (2+2)	13.8	8.2 (FYM- 3.5) 2.5 (VC: 1.3) 4.0- Pond treat.	N-21.5 P-12.2 K-13.3			
2.	Veg. waste	11.3	6.2 (VC-1.6) 6.5 As fodder	N-28.6 P-22.2 K-24.7	N=56.5 P=39.6 K=42.7	Total: Rs.4826/-	118 kg urea 247kg SSP 71.2 kg MOP
3.	Duck drop. (35)	1.21	As fish feed/silt	N-6.4 P-5.2 K-4.7			

Model at ICAR-RCER, Patna

Table 3: Recycling of farm waste and gain/saving of nutrients through 1 acre IFS model at ICAR-RCER, Patna

S.No.	Farm waste	Quantity produced (q)	Production/ usepattern (q)	Nutrient gain (kg)	Total Nutrient Gain Upon Recycling	Saving due to resource recycling (Rs.)	Fertiliser Saving (kg)
1.	Goat (20+ 1) droppings 24.9	18.5 (GM- 7.2) 6.4 (VC- 1.7)	N- 10.0 P- 5.8 K-11.6	N-44.0 P-29.5 K-31.2			93.0kg urea 184.0kg SSP 52.0kg MOP
2.	Veg. waste 66.2	18.4 (VC- 6.8) 50.0q- Asfodder	N- 14.1 P-10.2 K-14.8			3125	
3.	Poultry manure(600) 17.8	Used in crops (35.2)	N-20.7 P- 17.5				
4.	RWMML Straw 46.4	4.4 – mush. shed 1.6- Hut 42.8 q- sold	K- 9.6				

Employment Generation

It was often said there is hidden unemployment in Agriculture which can be best descried with an example with a seven members in a family (four adults and three children) , they possesses a land of one acre and rice- wheat cropping system is being followed. It was observed that all the members are engaged in the farming itself while it is not required. Most of the time they are sitting idlc. IFS model here is ready to overcome this type of problem by integrating more number of remunerative components in the system. By integrating one or two small components say 500 no. of poultry and 500 bags of mushroom cultivation, about 110 man days requirement is enhanced and these family members are able to perform these works easily. Women's and children can be also involved in many of agriculture based activities like nursery raising, mushroom cultivation, poultry farming, vermicomposting etc. Man power requirement under one acre IFS model (Crop + Goat + poultry + mushroom) and two acre model (crop + fish/duck + livestock) has been depicted in the figure 3. Man- power requirement by different combinations and component wise were also studied in the farmers' field and presented in table 4. It was observed that livestock base IFS model engaged the maximum no. of man- days and generated an additional 281man- days over cereal based cropping system only.

Table 4: Employment generation (man-days) by different IFS models

Farming Components	Crop	Hort.	Poult	Duck	Fish	Goat	Dairy	FYM/V. comp.	System emp.	Add. M-days
Crop alone (cereals)	237		-	-	-	-	-	-	237	-
Crop + Hort.	172	242						21	435	120
Crop/Hort. + fish + poultry	135	145	110	-	36	-	-	20	446	181
Crop/Hort. + fish + duckery	155	145	-	40	36	-	-	25	401	136
Crop/Hort. + fish + goat	135	145	-	-	36	110	-	30	456	191
Crop/Hort. (0.4ha)+fish + cattle	133	145	-	-	36	-	210	32	568	281
Crop/Hort. + fish + P +D	135	145	110	40	36	-	-	20	486	221
Crop(c/v)+ mush.+ goat	155	145	-	-	-	110	Mush . 40	20	470	205
Crop / Hort.+ P+Mush. + Goat (1 acre)	72	84	63	-	-	110	Mush 40	32	351	86
Crop / Hort. + Fish + D+ Cattle (2 acre)	133	145	-	40	36	-	210	32	619	382

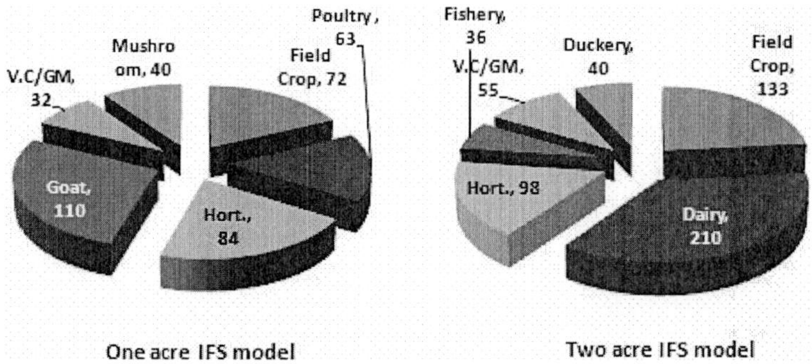

Fig. 3. Man- days requirement by different components under one and Two acre IFS models

Effect on soil fertility and water quality due to different integrations

Integrated farming system provides ample amount of organic wastes in form of organic manures after its recycling which contains abundant quantity of NPK and other micronutrients. It was observed that due to incorporation of organic manures in form of FYM + poultry droppings or goat + poultry droppings to IFS plots, available N, P & K increased over the plots in which only inorganic fertilizers were applied over a period of five years (Fig. 4). A clear cut decrease in organic carbon content in the soils was also observed which was fertilized with inorganic sources only while a slight increase was observed in respect of organic carbon content in the plots where organic sources were applied along with inorganic sources too. This finding advocated the utility of INM practices for getting an optimum yield level for a longer period of time. Nutrient content in vermicompost and duration for preparation of vermicompost from different sources varied in respect of material used, substrate used and enriching material used. It was found that when vermicompost was prepared from leaves then it took only 75-80 days and nutrient content was also higher over other sources. Cow dung was used for enriching the compostable materials. The details about other sources are presented in Fig.5.

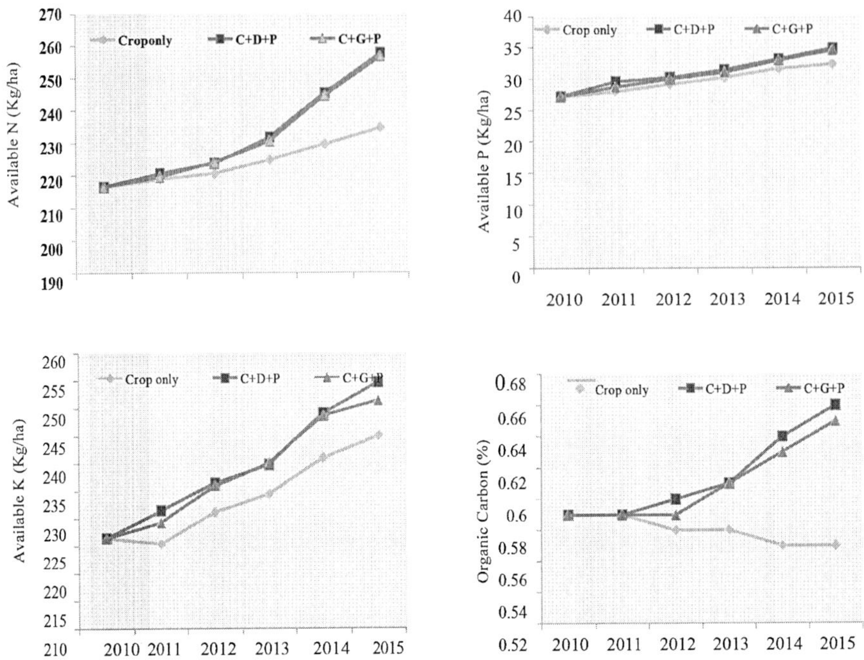

Fig. 4. Change in Fertility status due to nutrient recycling

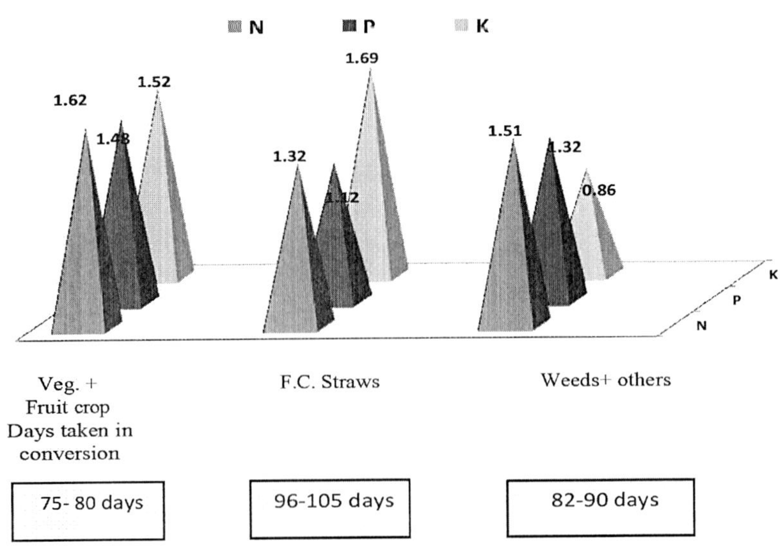

Fig. 5: Percent NPK in Vermicompost prepared from different sources

Duck- fish integration

Duck- fish integration is one of the promising integration for low lying areas where water is stagnated up to a period of 4-5 months in a year. For a pond area of 1000m^2 a total no. of 40-50 ducks can be incorporated and duck droppings may be used as feed source for the fishes (mix carp culture, 1000 fingerlings). There is no need to supplement with outside feed source if ducks were integrated. Some fodder like berseem/napier and banana leaves may be applied in the pond for fishes as surface feeder. It was observed that due to duck dropping, amount of dissolved oxygen decreases in the winter season because of slow decomposition rate of organic sources in winter while pH increases slightly during the winter season (Fig.6). It was also observed that the pond in which duks were integrated and allowed for swimming throughout the year, dissolved oxygen content was maintained to a optimum required level. Duck also act as natural weeder and they keep the pond clear by consuming the aquatic weeds and other things. From 40 ducks one can get about 6000 nos. of eggs/year and from a water area of 1000m^2 about 300-350 kg of carps can be harvested in a year which is highly profitable for a small farmer.

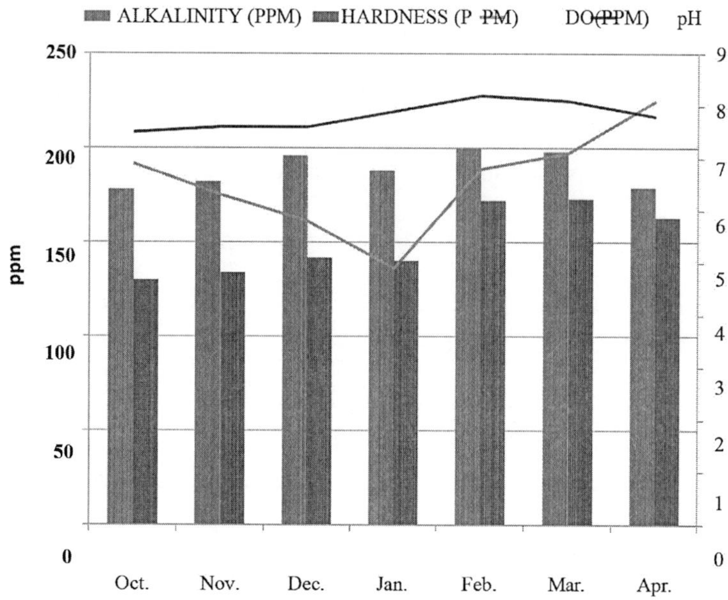

Fig. 6: Pond Water quality under Fish- duck integration

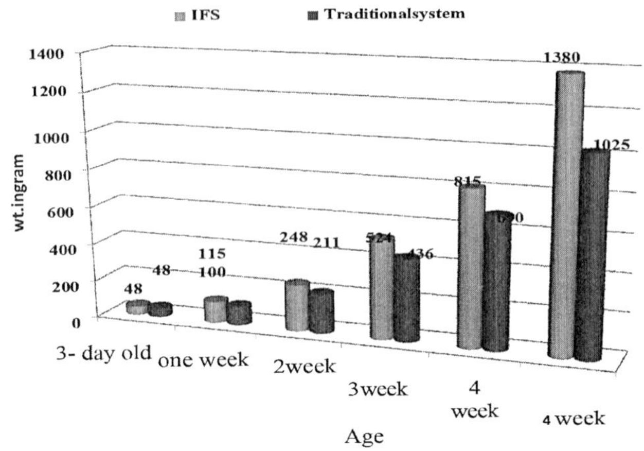

Fig.7. Av. Growth rate in poultry (IFS vs traditional system)

It was observed from Fig. 7 that average growth rate of birds was more in traditional system where commercial feed was fed to the birds upto the age of 5 weeks (1.4 kg) over the birds which were fed on the feed prepared by IFS produce like mixture of lentil, maize, rice bran, oil cakes etc. (1.02kg) but when we consider the profitability, IFS reared birds surpasses in terms of net returns (Rs.20-22/bird) over commercial fed birds (Rs. 10-12/bird) due to high cost of commercial feed over self made feed by using system produce. In contrast of body wt. of poultry it was observed that average body weight of goats (12 months age) were more by 1.3 kg under IFS over goat reared in traditional system (fig. 8). It may be due to the fact that the goats reared under IFS have got sufficient fodder throughout the year with balanced ration over traditionally fed goats that have faced fluctuated dietary supplements i.e. not balanced feed and care.

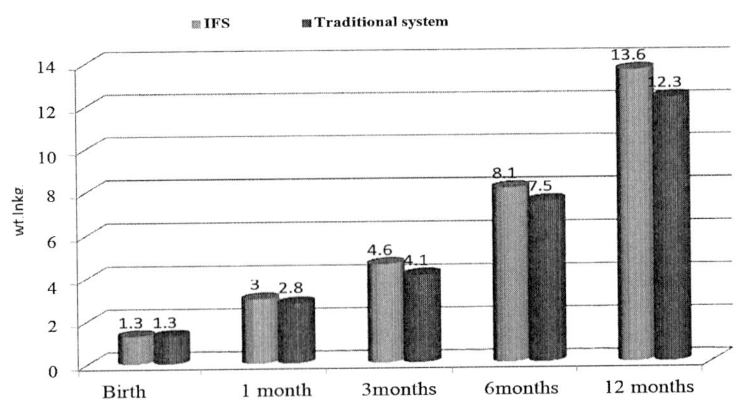

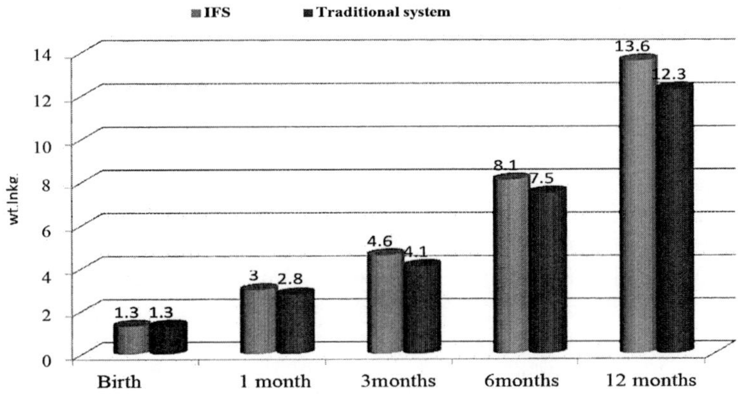

Fig. 8. Body wt. of Black Bengal goat at different age

Energy budgeting under IFS

Farming system is a resource management strategy to avail maximum efficiency of a particular system. Studies conducted at Goa revealed the higher energy use efficiency of IFS with rice. Integration of poultry and mushroom enterprises with rice-brinjal system required highest energy input whereas rice cropping alone recorded the least requirement of energy (Table 5). The energy output was maximum under rice-brinjal + mushroom + poultry. The output of multi-rice based enterprise was reasonably good varying from100.91 to105.63 MJ/ ha. It is thus evident that efficient utilization of scarce and costly resource is the need of the hour and can be accrued by following the concept of IFS through supplementation of allied agro-enterprises (Korikanthimath and Manjunath, 2009).

It was observed that under one acre IFS model (Crop + Goat + poultry + mushroom), developed at ICAR Research Complex for Eastern Region, Patna total input and output energy was calculated as 68, 491 MJ and 2,17,548 MJ respectively (Input/output ratio:: 1: 3.2) while under two acre IFS model (crop + fish/duck + livestock) was reported as 296709MJ and 7,52,415 MJ, respectively (Input/output ratio:: 1: 2.5), means crop + goat + poultry + mushroom integration is more energy efficient over crop + fish/duck+ livestock integration (Fig.9).

Table 5. Energy (x10^3 MJ/ha) budgeting for rice based integrated farming systems

Integrated farming system	Pooled mean energy		
	Total input	Total output	Efficiency
Rice-fallow	11.56	78.18	6.76
Rice-groundnut+mushroom+poultry	46.08	102.86	2.24
Rice-cowpea+mushroom+poultry	43.79	105.63	2.41
Rice-brinjal+mushroom+poultry	52.03	165.33	3.18
Rice-sunhemp+mushroom+poultry	41.44	100.91	2.44

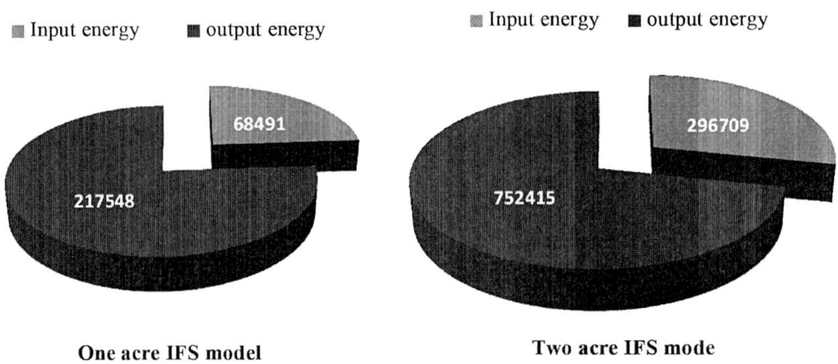

Fig. 9. Total input & output energy in MJ

Income sustainability index

Income sustainability index gives an idea about sustainable income from an IFS model for a longer period of time for its existence. It is assumed that for better sustainability of the IFS model sustainability index should be more than 0.5. Developed IFS models were evaluated based on sustainability index (SI) as described by Vittal *et al.* (2002). The SI for any IFS model can be computed as : *SI= (NR- SD) / (MNR)* where, NR stands for net returns obtained under any model, SD stands for standard deviation of net returns of all models and MNR stands for maximum net returns attained under any model. A suitable and viable IFS model could be identified for their existence based on net return, sustainability index, employment generation and improvement in soil fertility attained over a period of time. Here, among different IFS model evaluated, Crop +goat +poultry + mushroom integration provided highest ISI (79. 2) and was followed by Crop + goat + fish + duck (77.2) which was again followed by Crop = goat + poultry (75.7) and vegetable + goat integration (75.1) whereas, income sustainability index (ISI) from rice- wheat system was in negative (Table 6).

Table 6: Annual net return and Income Sustainability index of different farming systems (in ha)

IFS components	RGEY (t)	Gross Income (Rs.)	Net Income (Rs.)	B: Cratio	Man-days	I.S.I.
Rice - wheat	8.1	105650	30534	1.4	237	-0.037
Crop + goat	13.5	189000	77449	1.7	322	28.9
Crop + cattle	15.5	201350	72950	1.5	405	21.3
Veg. + goat	19.0	254845	139414	1.8	452	75.1
Hort. + Cattle	19.9	259690	116585	1.8	535	58.7
Crop + poultry	18.4	238985	96425	1.7	390	44.9
Crop + goat+ fish	18.9	245720	99524	1.7	333	47.0
Crop +goat + fish + duck	23.6	306840	142214	1.8	398	77.2
Crop + cattle + fish	16.2	211030	77030	1.6	435	31.7
Crop + cattle + fish + duck	18.5	242530	100530	1.7	470	47.7
Crop + goat + poultry	22.8	296445	140614	1.8	416	75.7
Crop+goat +poultry+ mush.	24.5	318240	144315	1.9	439	79.8
S. D.	18.6	43247.9	30588.7	0.14	64.4	-
C.V.	10.8	17.8	29.6	8.5	15.4	-

Research experiences on IFS Farmers' fields

While selection of farmers following points were taken into consideration viz.

i) Technologies must match with farmers' needs and resources for a better livelihood,

ii) Solves their problems

iii) Prioritization of problems capabilities in adopting technologies

iv) Economic status of the farmers

v) Market, social structure, indigenous knowledge and policy of the Govt.

Before demonstration of IFS model, focus group discussion was conducted that included all section of the society including farmwomen and weaker sections, water user associations and other existing institutions. The team further conducted participatory need assessment to know the inherent nature of problem of the area. The facts were documented and the problems were analysed. The problems were prioritized for addressing. The relevant solutions recommended by various research organization and other stakeholders were also collected and discussed with wider participation of farm families to match the possible solutions.

Out of the six selected districts, in Patna, rice is the main crop. It accounts for more than one third gross area sown. Other important food grains grown are maize, pulses and wheat. Non-food crops consist mostly of oil-seeds, cash crops such as vegetables, water- melons etc. are also grown in diara belt of the district. The land in the district is too fertile to be left for wild growth. The district has a population density of 1,803 persons per square kilometre and sex ratio of 892. The literacy rate of the district is 72.47%. Agriculture is the main source of occupation in Nalanda district. The farmers mainly grow paddy, besides potato, and onion. Few people of the district are also involved in handloom weaving.

Darbhanga is flood prone district. Prolonged land submergence compels farmers to take photosensitive tall rice varieties with poor productivity (1.07 t/ha) as well as late sown wheat (1.8 t/ha). A total of 78.5 per cent houses are *kaccha* and 66.3 per cent population is below poverty line. High population density and low availability of food grains (65 kg/head/yr) reflect on poor health of the population. However, stagnant surface water and ground water support a high yield of boro rice. Winter crops like maize, wheat and pulses are also cultivated. Aquatic crops like *Makhana* and *Singhara* are cultivated in significant area. The livelihood is based on agriculture and the income generated by migrant labourers, working in different parts of the country and abroad.

In Samastipur, rice-wheat is the major cropping system with higher variability in yield of maize, vegetables, fruits and fisheries. Sugarcane, potato and maize (seed production)and spices are the other cash crops produced in the district. A strong crop-livestock interaction also exists. Population density is high with a population density of 1175 per km^2. Crop, livestock and off-farm employment are the major sources of income. The total productivity of rice- wheat system is 3.2 t/ ha. Food scarcity and employment opportunities are major challenges where only 33per cent area is irrigated and large area is flood affected.

In Vaishali district, rice-wheat is the main cropping system, followed by fruits and vegetables. Poor productivity of rice (1.05 t/ha) and wheat (2.2 t/ha) is mainly due to poor access to the available water resources (economic scarcity) with rainfed agriculture in about 61.5% of the area, flooding, water logging and soil sodicity etc. Very high population density (1332/km^2) with 63 per cent population below poverty line is supported by 5.6 lakh cattle population. The milk availability is only 49 kg/head/yr. The livelihood is mainly supported by crops, livestock, and some remittances by the migrant laborers.

In Munger, plains, *diara (riverine lands)* and plateau areas exist in close vicinities. The rice-wheat system in the rainfed area (57%) with 3.5 t/ha productivity, coupled with 3.4 lakhs livestock population and off-farm activities of the district are major sources of livelihood. The large *diara* lands have poor productivity but high potential of winter maize and summer vegetable production.

Based on PRA, available resources, farmers' preferences & need and economic viability of the components, social acceptance etc. following integrations have been undertaken for each half acre, one acre, two acre IFS models and IFS models for wetland and were evaluated at the 155 farmers' fields at Nalanda, Vaishali, Munger, Samastipur, Darbhanga and Patna district of Bihar covering an area of 93.0 ha (Table 7,8&9).

Major integrations under integrated farming System

A. Half Acre Model

- Crop (Rice- wheat) + goat(10)
- Crop (Rice- wheat/maize) +goat (10) +backyard poultry (200birds/batch)
- Crop (Rice- wheat/maize) + buffalo (2) + backyard fishery (200m^2)
- Vegetables + goat(10)
- Crop (Rice- wheat) + vegetable +mushroom

B. One Acre Model

- Field crop + Vegetable + goat (20+1)
- Field crop + vegetable + poultry (200 x6)
- Field crop+ fruits + Cow(2+1)
- Crop + goat (20 +1) + Poultry +mushroom
- Field crop+ vegetable + mushroom (600bags)

C. Two acre model

- Field crop + Dairy (2)+ fish (1000m^2)
- Field crop + Dairy (2) + fish (1000 m^2) + Duck(35)
- Field crop+ goat +Fish
- Crop + goat + Poultry +mushroom
- Field crop+ Orchard +Dairy +mushroom

D. IFS model for wetland

- Rice +fish
- Fish + fruit +vegetable
- Horticulture +fish
- Horticulture + fish +duck
- Makhana + fish – water chestnut
- Makhana + fish +horticulture
- Fish + fodder + livestock + goat +duck

- Integrated fish culture +livestock
- Crop +tree

Among different IFS models evaluated at 155 farmers' field of Patna, Nalanda, Munger, Vaishali, Samastipur and Darbhanga, crop + goat + backyard poultry integration resulted into average maximum annual net returns (Rs. 52, 650) with half acre IFS model (Table 5), Crop + goat + poultry + mushroom (Rs.82,000 and Rs. 1,56, 570) under one and two acre IFS models (Table 7 & 8), respectively with ample addition of N, P and K in the soil due to recycling of nutrients within the system.

- Sustainability index of above integrations were also found higher 0.71, 0.73 and 0.65 for half, one and two acre IFS models, respectively. Animals were kept on restricted grazing under half acre IFS model while 10 percent and 20 percent of total land area were was found optimum for year round fodder production for goat and dairy component integration, respectively under one and two acre IFS models.
- Integration of animal/ vegetable/ birds increased the annual returns by 2-3 folds over traditional rice- wheat cropping system at farmers' fields.
- No additional feed were given to fish as feed. Pond was supplemented with cow dung (2.2 q), duck droppings (6.2q) and banana, berseem, vegetable leaves etc. Mix carp culture (Katla/ Grass Carp, Rohu and Mrigal in the ratio of 3:2:5) were reared in the pond to avoid competition for feed. Duck droppings were allowed to go into the ponds directly which had supplemented feed for fishes and fertility buildup of pond's soil.

Table 7: Annual net returns and Sustainability index of different half acre Integrated farming systems at Farmers' field (Mean value of the year 2009-14)

IFSmodels	RGEY(t)	Production cost (Rs)	Gross returns (Rs)	Net returns (Rs)	Nutrient addition in the system through recycling(Kg)	Employment generation (man-days)	Income sustainability index
Rice-wheat	1.6	8,850	19,200	10,350	——	58	-0.09
Crop + goat	5.2	25,750	62,400	36,650	N-22.6 P-18.4 K-17.8	120	0.41
Crop + goat (10) + backyard poultry (200 x 6cycles)	10.7	75,750	1,28,400	52,650	N-29.2 P-25.8 K- 26.2	165	0.71
Crop + buffalo (1+1) +backyard fishery (400 m^2)	5.0	35,420	60,000	24,580	N-19.2 P-18.6 K-17.2	150	0.18
Crop + Vegetable+ goat	6.8	37,150	81,600	44,450	N-27.5 P-23.7 K- 24.1	180	0.56
Crop+ Vegetable+mushroom	5.8	29,150	69,600	40,450	N-22.9 P-18.2 K-19.6	120	0.48

Note: RGEY: Rice grain equivalent yield

Table 8: Annual net returns and Sustainability index of different one acre Integrated farming systems at Farmers' field (Mean value of the year 2009-14)

IFSmodels	RGE Y (t)	Production cost (Rs)	Gross returns (Rs)	Net return s (Rs)	Nutrient addition in the system through recycling g(Kg)	Employment generation (man-days)	Income sustainability y index
Rice- wheat	3.5	17,800	42,000	24,200	-	115	-0.04
Field crop+ Vegetables+ goat (20+1)	13.1	71,250	1,57,200	75,950	N-28.4 P-26.1 K-26.7	223	0.54
Field crop+vegetables+ poultry (200 x6cycles)	13.9	77,985	1,56,800	78,815	N-51.2 P-38.3 K- 36.8	196	0.62
Field crop+ fruits + Cow (2+1)	13.3	75,825	1,49,600	73,775	N-31.7 P-27.2 K-26.8	286	0.50
Field crop+Vegetable+ goat (20+1) +Poultry (200 x 6) +mushroom	18.6	1,11,200	1,93,200	82,000	N-63.8 P-45.7 K- 43.2	287	0.73
Field crop+ veg. + mushroom (600 bags)	10.7	43,265	1,18,400	75,135	N-20.8 P-17.1 K-13.2	167	0.60

Note: Field crop = field crops + fodder

Table 9. Annual net returns and Sustainability index of different two acre Integrated farming systems at Farmers' field (Mean value of the year 2009-14)

IFSmodels	RGEY (t)	Production cost (Rs)	Gross returns (Rs)	Net returns (Rs)	Nutrient addition in the system through recycling (Kg)	Employment generation (man-days)	Income sustainability index
Rice- wheat	6.8	35,150	81,600	46,450	——	217	0.02
Field crop+ Dairy (2)+ fish (1000m^2)	16.4	92,750	1,96,800	1,04,050	N-38.4 P-35.3 K-31.8	436	0.35
Field crop+ Dairy (2)+ Fish (1000m^2)+ Duck (35)	19.1	99,810	2,29,200	1,29,390	N-49.2 P-43.2 K-36.7	492	0.49
Field crop + goat+ Fish	17.2	80,170	2,06,400	1,26,230	N-57.5 P-53.3 K-54.2	359	0.47
Crops + goat + Poultry + mushroom	28.7	1,07,830	2,64,400	1,56,570	N-65.4 P-58.3 K-59.1	502	0.65
Field crop+ Orchard+Dairy + mushroom	19.3	1,02,110	2,31,600	1,29,490	N-61.3 P-52.8 K-50.1	484	0.49

Note: Field crop = field crops + fodder

Fig. 10. Integration of crop + livestock+ fishery in two acre at Village Yashwantpur, Nalanda (Nalanda, Bihar)

Fields (Wetland Ecosystem)

In Bihar about 9.0 lakh ha area falls under wetland where water stagnates from 4 months to 6-7 months ranging from 60 cm to more than 3.0 m. These wetlands are one of the prime resources for inland fishery besides production of agricultural crops. A systematic and integrated approach was applied to improve the soil and water productivity of these areas. Up scaling of IFS models for wetlands was done in Canal area of Patna district, chaur area of Samastipur and Vaishali district and wet lands of Darbhanga districts of Bihar. Such lands were targeted for up scaling through the IFS technologies in participatory approach Kumar *et al*, 2015). The models were developed through individual farmers and SHGs. An intervention on fished based farming system (fish + livestock+ duck + Hort.) has been demonstrated in 40 ha chaur areas at Sarairanjan of Samastipur district of Bihar that resulted net average income of Rs. 1.24 lakhs/acre (around 5.5 times more) besides an additional employment generation of 160 days over rice- wheat cropping system (Table 10 &11).

Table 10: Annual net return and employment generation through fish based IFS at Farmers' field (1.0 acre), Sarairanjan, Samastipur (Mean value of the year 2009-12)

IFS models	RGEY(t)	Production cost (Rs)	Gross return (Rs)	Net returns (Rs)	Employment generation (man-days)
Rice-wheat	3.2	16650	38400	21750	110
Fish (2000 m^2) + Fodder (1000m^2) livestock (2+1) Goat (5) + Duck (50)+ Horticulture on pond's bund	21.8	1,37,520	261600	82,080	270

Fig.11. Aquaculture based IFS at Sarairanjan (Samastipur, Bihar)

In canal command of seasonally waterlogged area of Patna district, integration of fish in sunken trenches with horticulture on raised beds gave net average income of Rs. 46,250/- per ha and an additional employment of 45 mandays over rice fallow system (87 man days) was generated (Table 11).

Table 11: Relative returns of different multiple use system in canal command, Patna, Bihar

S.No.	Multiple use system	Gross income (Rs/ha)	Net income (Rs/ha)	B:C ratio	Employment (man/days/ha/yr)
1.	Rice + fallow	46,600	26,500	2.32	87
2.	Rice -fish in waterlogged area	63,000	35,900(35)*	2.25	112
3.	Fish in sunken trenches, horticulture on raised beds	89,250	46,250 (75)*	2.76	132
4.	Secondary reservoir + horticulture & vegetable on dykes	1,11,000	56,420 (113)*	2.34	148

* Value indicate % increase over traditional rice production

Makhanaonly Rice-Makhana Makhana-fish

Fig.12: Integration of Makhana with rice and fish at Darbhanga

Rice- fish culture

Paddy cum fish culture is an age old tradition in rainfed as well as irrigated ecosystem of Eastern regions. Fishes are cultivated in paddy fields either simultaneously or in rotation. Both freshwater and saline fields can be utilized for the purpose. Commoncarp (*Cyprinuscarpio*) is the most preferred species for paddy-fish culture. However,Catla (*Catlacatla*), Rohu (*Labeorohita*), Mrigal (*Cirrhinusmrigala*), Bata (*Labeobata*), Java punti (*Puntiusjavanicus*), Silver carp (*Hypophthalmichthys molitrix*) and giant freshwater prawn (*Macrobrachium rosenbergii*) are some other suitable species for paddy-fish culture. Sea bass (*Latescalcarifer*), Mullets (*Liza parsia, L. tade* and *Mugilcephalus*), Catfish (*Mystusgulio*) and Tiger prawn (*Penaeusmonodon*) could also be reared in fields with saline waters. Fish production @ 500-800 kg/ha from freshwater paddy fields and 400-1500 kg/ha from saline water fields is expected in six months.

Fig. 13: Rice- Fish Culture

Fig. 14: Cage Fish Culture

Enhancing productivity in chaur land

The various technologies were demonstrated in Kaincha Chaur, Sakroli Bucholi village, Jandaha Block, Vaishali district, Bihar. Following interventions were adopted:

Fish culture in cages

Floating cages of 3 x3 x 2 m size were installed to grow fry to fingerlings. *Catlacatla* and *Labeorohita* have attained 11.5cm/20g and 13cm/35g respectively in 105 days in cages with a survival rate of 80% at a stocking density of 47 fry / m^2 during 2008-09. Three crops per year could be obtained by farmers in cage rearing with net profit of Rs 3000 per year per cage.

Horticulture -Vegetable and livestock production in fish pond system

Fig. 15. Low cost duck shed

There are diversified systems for the management of wetlands, which combine the norms of capture and culture based fisheries. In this system, the marginal areas of wetlands are cordoned off for culture fishery either in ponds or in pens and the central portion of wetland is left for capture fisheries. To enhance the water productivity, two ponds were reconstructed during 2008-09. Cultivation of vegetables (Bottle Gourd) on the embankment yielded 20.0 t/ha. Poly culture of indigenous major corps and exotic carps were conducted in ponds. A duck house was utilized for keeping 22 Khakhi Cambell in the system. The eggs produced (52500 per ha) through integration of ducks yielded an additional income of Rs.2, 62,800/-ha. The fish yield ranged between 3.5-3.6t/ha.

Low cost portable hatchery for raising fish seed

Fig. 16: Portable hatchery

Portable FRP Carp Hatchery for raising fish seeds was established. During 2009-10, the successful operation of hatchery resulted in the production of 80 lakhs spawn of major carps. This hatchery is useful in producing fish seed at farmer's field. In one run, it has the capacity to produce 1.0-1.2 million spawn. Such a low costing hatchery can easily beprocured by group of farmers. Farmers were advised to make SHG and by pooling their resources, the hatchery can be procured. This offers scope for its wide adoption in rural India. This will definitely enhance the fish seed availability.

Fish–Livestock and Fodder integration in seasonally waterlogged area

The possibility of pasture-goat integration was also explored in seasonally waterlogged area of Patna district. Waterlogged area (0.3 ha) with water depth of 0.5-1.0 m for 3-5 months was converted into alternate trenches and beds for

Fig. 17. Pasture-goat integration

pasture production. The possibility of pasture-goat integration was also explored in seasonally waterlogged area of Patna district. Waterlogged area (0.3 ha) with water depth of 0.5-1.0 m for 3-5 months was converted into alternate trenches and beds for pasture production. Trenches of 1.5- 2.0 m depth was prepared and excavated soil was used for bed preparation for pasture production. On the bed leguminous and non- leguminous grasses were planted. Fifteen Black Bengal breed of goats were integrated with the system and in trenches (0.08 ha) fish yearlings (10000 per ha) were stocked. About 10.0 t/ha fodders were produced with meat production of 90 Kg/ha/yr. A total of 1.7 t/ha fish from the water area of trenches was obtained.

Makhana – Fish – Water chestnut integration

In pond based ecosystem of Darbhanga district, analysis of results from total water bodies of 50 ha (96 beneficiaries) exhibited fish, Makhana, and singhara yield in the range of 0.18 to 0.4, 1.06 to 2.06, and 3.08 to 8.8 t/ha, respectively. In makhana+fish–water chestnut integration, makhana as a primary crop showed a net profit of Rs. 20,015/- per ha with an employment generation of 240 man days/ha/year (Table 6). The fish as a secondary crop integrated in makhana ponds showed an additional net income of Rs 11,806/= per ha with an employment generation of 24 man days/ha/year, whereas singhara, taken as tertiary crop, generated an additional net income of Rs13,445/= per ha with an employment generation of 83 man-days/ha/yr (Table 12). Thus, integrated system resulted in a benefit cost ratio of 1.94 as against 1.63 for the traditional system (makhanaonly).

Table 12: Economics of integrated system of makhana cultivation

	Traditional system	Integrated System		
	Makhana only	Makhana	Fish	Water chestnut
Productivity (t/ha)	1.40	1.33	0.27	3.7
Cost (Rs./ha)	32, 693	31, 058	11, 323	13, 445
Net Income(Rs./ha)	20, 614	20, 015	11, 806	20, 614
Labour employed (man days/ha)	247	240	24	83
Benefit cost ratio	1.63		1.94	

Traditional system (Makhanaalone) Rice –fish integration

Makhana – fish integration at Fish in trenches Horticulture on ponddykes

Fig. 18. Birpur & Beeranchak villages in Patna district

Makhana + fish + horticulture on pond dykes integration at Darbhanga district resulted in the average yield of makhana and fishes as 1.2 and 0.31 t/ha, respectively. The gross and net return from the system, including horticultural crops, were estimated as Rs 1, 19,246 and Rs 50,439/ha, respectively.

Restoration of Waterlogged/marshy Areas through Agroforestry Interventions

Agroforestry land use is very vital for rehabilitating waterlogged/marshy and degradedlands. Raised bed method could be adopted in marshy areas for planting ofsaplings. Species like *Anthocephalus chinensis*, *Casuarina equisitifolia*, *Dalbergiasissoo*, *Eucalyptus* spp., *Ficus* spp., *Mangifera indica*, *Melia azadirachta*, *Moringa* spp., *Morus* spp., *Musa paradisiaca*, *Populus deltoides*, *Salix* spp., *Syzygiumcumini*, *Terminalia arjuna*, *T. bellirica*, *T. tomentosa* and *Vitexnogundo* etc have been found suitable for cultivation in such areas besides various bamboo species. Such lands could also be rehabilitated with land configuration and adopting pisciculture and agroforestry together.

Fig. 19: Tree and hedge plantations for restoration of water logged area

Integrated Fish Farming

Integrated farming of fish and livestock is vital technological intervention in flood prone ecosystem. The aim of integrated fish farming (IFS) is the recycling of animal wastes to serve as fertilizers, and sometimes as food for fish raised in ponds, enclosures and cages and thereby minimize the cost of cultivation. Grow-out ponds with1.5- 2.0 m water depth and having good water holding capacity are ideal for IFS. About 8000 to 10,000 fish fingerlings could be stocked in 1 ha of pond. However, stocking of yearlings could be preferred to achieve higher growth rate and yield. To fertilize one ha of pond, the stocking density of livestock / birds has been worked out to be 5, 450, 500, 30 and 55 nos. of cattle, poultry, duck, pig and goat, respectively (Table 11).

Fig. 20. Integrated Fish – livestock farming

Summary

Integrated farming system approach advocates a change in the farming techniques for maximum production in the cropping pattern and takes care of optimal utilization of resources. The farm wastes are better recycled for productive purpose in the integrated system. Farm as a unit is to be considered and planned for effective integration of all enterprises. The future of Indian agriculture depends heavily on the development of appropriate farming system as it fits to resource poor farm families of different agro ecological zones. The endowment of abundant sunshine, long growing season, responsive soil type and combination of surface water, ground water and seasonal rains and above all a progressive peasantry offer vast scope for an intensive farming system through multiple cropping and diversified farming including animal husbandry, forestry, sericulture, fisheries and the like. The challenge is to upgrade technological and social disciplines on a continuous basis and integrate these disciplines to suit the region and the farm families in a manner that may ensure increased production with stability, ecological sustainability, nutritional security and equitability.

Future Thrust

- Creation of database on Integrated farming systems throughout the country in relation to type and size of integrated farming systems, enterprise selected and their way of allocation, infrastructure, economics, economic sustainability of the system etc. under different ecological situations.
- Development of ecologically stable, environmentally sound and location specific low cost viable IFS modules for different holding sizes which are socially acceptable is required.
- On –farm testing and refinement of the developed modules according to the farmers, need and requirement as it is a continuous process i.e. addition of profitable components and replacement of less profitable components with time, choice of the farmers and availability of market.
- Need to study the sustainability of the developed or identified farming systems under different agro-climatic situations in the long run including high value crops.
- Need to study the nutrient dynamics of soil, accumulation of carbon and carbon sequestration with continuous cropping and recycling of organic resources in form of plant or animal wastes with different systems overtime.
- Need to identify the constraints in adoption of identified integrated farming systems for particular area or locality.
- Development of indigenous technology know how (ITK) of IFS existing in the farming community and its scientific validation.
- Need to prepare a strong policy draft for the consideration of planners for its promotion and creating awareness at large scale with some pity financial assistance either through loans or subsidy.

From the above studies and experiences it was emphasised that farming in integrated mode is highly remunerative for the farmers on one hand while on the other hand it also accelerates soil sustainability and ensures income and nutritional security of a farm family by providing high value enterprises, protein, fat and carbohydrates. Addition of organic residues in the form of animal and plant wastes help in improving the soil – health and thereby productivity over a longer period of time with lesser environmental hazards with increased profit margin. System comprising of crop components, dairy, poultry and fishery is the most remunerative and efficient as it provided the highest system productivity for irrigated agro-ecosystem of North eastern plain zone while suitable IFS model for Indian Central Himalaya region is fishery + poultry + vegetable farming which has considerable potential to provide food security, nutritional benefits, employment generation and providing additional income to resource poor small farmers. In general, IFS enable the agricultural production system sustainable, profitable (3-6 fold) and productive on long term. About 90-95 percent of nutritional requirement is self-sustained through resource recycling which curtails the cost of cultivation and increases profit margins and employment. Therefore it is imperative to state that to sustain food and nutritional security, IFS approach is promising and will conserve the resource base through efficient recycling of residues and wastes within the system. IFS models developed for different ecological ecosystems and sub systems can be tuned through farmers' participatory trials with multilevel interventions itself on the farmers' fields.

References

Ganesan, S., Chinnaswami, K.N., Balasubramaniam, A. and Manicksundaram, P. (1999). Studies on rice based farming systems with duck-cum fish culture in deltaic region in Thanjavur district of Tamil Nadu. *Farming Systems News Letter*. 1(2), 15.

Gill, M. S., Singh, J. P. and Gangwar, K. S. (2009). Integrated Farming System and agriculture sustainability. *Indian Journal of Agronomy* 54 (2), 128-139.

Inman,D., Khosla, R., Westfall, D.G., Reich, R. (2005). Nitrogen uptake across site- specific management zones in irrigated corn production systems. *Agronomy Journal* 97: 169-176.

Sanjeev Kumar, B P Bhatt, ADey, Ujjwal Kumar, N Chandra and Shivani (2015). Sustainable Integrated Farming System Models for small and marginal farm holders of Bihar. In: Research in Farming Systems (Edts. B. Gangwar, J P Singh, A K Prusty and Kamta Prasad). Todays &Tommorow's Printers and Publishers, New Delhi, pp- 229-249

Kumar Sanjeev, Subhash, N., Shivani, Singh, S. S. and Dey, A. (2012). Evaluation of different components under Integrated Farming System (IFS) for small and marginal farmers under semi-humid climatic environment. *Experimental Agriculture*, 48 (3): 399-413.

Kumar Sanjeev, Singh, S. S. Meena, M. K. Shivani, S., Dey, A. (2012). Resource recycling and their management under Integrated farming system for lowlands of Bihar. *Indian Journal of Agricultural Sciences* 82(6):504-510.

Korikanthimath, V.S. and Manjunath, B.L. (2005). Resource use efficiency in integrated farming systems. (in) Proceeding of Symposium on Alternative Farming Systems: Enhanced income and employment generation options for small and marginal farmers pp. 109–18 PDCSR,Modipuram.

Kuppusamy, M. (1989). Feasibility of introducing sericulture as a component in the farming system in Thanjavur district. M.Sc (Ag.). Thesis, Tamil Nadu Agrl. Univ., Coimbatore.

Porpavai, K. A. Sheit Dawood, N. Kampuchetty and T.B. Ranganthan. 2000. *Madras Agric. J.*, 87 (4-6): 361-363.

Radhamani, S. (2001). Sustainable integrated farming system for dryland vertisol areas of Western Zone of Tamil Nadu. Ph.D.(diss) Tamil Nadu Agricultural University, Coimbatore.

Rajkumar, S. (1988). Feasibility of introducing dairy as complementary enterprise for small farmer in Thanjavurdelta. M.Sc.(Ag). Thesis, TNAU, Coimbatore.

Sheik Dawood, A. P.Santhi, K.Ponnusamy and P.Muthukrishnan. (1996). Integrated Farming System for lowland and cauverydalta zone. *Farming Systems* 13:(3-4):11-14.

Sivamurugan, 2001. Sustainable Farming system under irrigated upland situation. Ph.D.(thesis) Tamil Nadu Agricultural University, Coimbatore.

Vittal, K.P.R., MauthiSankar, G. R., Singh, H. P. and Sharma, J. S. (2002). Sustainability index. In: Sustainability of practices of Dryland Agriculture: Methodology and Assessment, Central Research Institute for Dryland Agriculture, Hyderabad.